S0-BFD-083

APPLIED KINEMATICS

Arthur J. Ramous

Associate Professor
Norwalk State Technical College
Norwalk, Connecticut

Prentice-Hall, Inc.

Englewood Cliffs, New Jersey

© 1972 by
PRENTICE-HALL, INC.
Englewood Cliffs, New Jersey

All rights reserved. No part of this
book may be reproduced in any form
or by any means without permission
in writing from the publisher.

Current printing (last digit):
10 9 8 7 6 5 4 3 2 1

13-041202-3
Library of Congress Catalog Card Number: 73-150393
Printed in the United States of America

PRENTICE-HALL INTERNATIONAL, INC., London
PRENTICE-HALL OF AUSTRALIA PTY. LTD., Sydney
PRENTICE-HALL OF CANADA, LTD., Toronto
PRENTICE-HALL OF INDIA PRIVATE LIMITED, New Delhi
PRENTICE-HALL OF JAPAN, INC., Tokyo

To

OLGA, DONNA, ARTHUR AND JOHN

PREFACE

This book is an introductory text written for students studying Kinematics or Mechanisms, in technical institutes, technical colleges, community colleges, four-year technical programs and industrial programs. No previous knowledge of physics or calculus is required.

There are numerous figures, photos and a large number of examples in this text. These examples have been carefully selected to emphasize the theory previously discussed or to broaden the theory. To reduce confusion each example has a "Given" and "Determine" statement. In many examples a figure has been presented along with the given statement so that there should be little doubt in the reader's mind as to what information is initially presented. In many examples the solution is presented in a sequence of numbered steps.

Great importance has been placed upon the problems at the end of each chapter. Many are from actual engineering designs. The problems were intentionally varied in difficulty so that the instructor has some means of adjusting the level of the course. A four-year technical school may lean toward the difficult problems, whereas a two-year school may use the "average" type problems. Some of the lengthy problems can be used in a board oriented course.

To obtain maximum use from this text the student must actively participate in the examples and problems.

Many instructors feel that a terminal or short course on Mechanisms should exclude acceleration. Therefore, the chapters which follow acceleration (Chapters 6, 7, 8) have been written with a minimum of reference to acceleration. This allows those instructors, who wish, to skip Chapter 5 (Acceleration).

The author wishes to thank Mr. Theodore Watkin for his review of Chapter 7 (Gears) and for his many helpful suggestions. Appreciation is expressed to Mr. Paul Czerepacha and Mr. Powell Lincoln for their assistance in photographing several of the mechanisms which appear in this text. Appreciation is also expressed to the many companies that provided technical data and photographs. Special appreciation goes to Pitney-Bowes, Inc. for their technical assistance and for the use of their copying facilities. Finally, but not least, the author wishes to thank his wife, Olga, for the many hours she spent typing the manuscript.

The following companies and organizations provided technical assistance:

Addressograph Multigraph Corp.
American Cam Co.
American Gear and Manufacturing Association
American Stock Gear Div. of Perfection American, Inc.
Ametek/Testing Equipment
Aronson Machine Company, Inc.
Atlas Electric Devices Co.
Audio-Visual Research
Automat Precision Engineering Ltd.
Avco Lycoming Div.
Baldwin-Lima-Hamilton Corp.
Beloit Corp.
Boston Gear Works
Briggs and Stratton Corp.
Bristol Motors
Brown and Sharpe Manufacturing Co.
Burndy Corp.
Century Projector Corp.
Clark Equipment Co.
Colt Industries, Pratt and Whitney, Inc.
Commercial Cam and Machine Co.
Deere and Co.
Delta Corp.
Design News
De-Sta-Co. Corp.
Detroit Diesel Engine Div. of General Motors Corp.
Dodge Div. of Chrysler Motors Corp.
Dravo Corp.
Energy Conversion Systems Corp.
Eonic Inc.
Erie Strayer Co.
E. W. Bliss Co.
Ex-Cell-O Corp.
Fairchild Hiller Corp.
Fairfield Manufacturing Co.
Ferguson Machine Co.
Ford Motor Co.

Gast Manufacturing Corp.
Geartronics Corp.
General Electric Co.
General Radio Co.
Gleason Works
Haydon Switch and Instrument, Inc.
Ideal Industries, Inc.
Industrial Press, Inc.
Jabcobsen Manufacturing Co.
Jones and Lamson Div. of Waterbury Farrel & Textron Co.
Keuffel and Esser Co.
Link-Belt Div. of FMC Corp.
L. R. Nelson Mfg. Company, Inc.
Machine Design
Malleable Founders Society
Matsushita Electric Industrial Company, Ltd.
Mattel, Inc.
McGraw Hill, Inc.
Metron Instruments, Inc.
Montgomery Ward
Moore Special Tool Company, Inc.
Mower Tiller Wholesale
NASA
Outboard Marine
Philadelphia Gear Corp.
Pitman Manufacturing Co.
Pitney-Bowes, Inc.
Plessey Airborne Corp.
Portable Electric Tools, Inc.
Power Instruments Inc.
Product Engineering
Remco Industries, Inc.
Roto Broil Corp.
Rotork Engineering Company, Ltd.
Sier Bath Gear Company, Inc.
Sikorsky Aircraft Div. of United Aircraft Corp.
Spiroid Div. of Illinois Tool Works Inc.
Sunbeam Corp.
The Cincinnati Gear Co.
The Fellows Gear Shaper Co.
The Randolph Co.
Timber-Top-Inc.
Turner Uni-Drive Co.
Twin Disc, Inc.

United States Army Tank Automotive Center
Wallace Automation, Inc.
Wildman Jacquard Co.
Winfred M. Breg, Inc.
Winsmith Div. of UMC Industries, Inc.
Woodlands Div. of Beloit Corp.

A. J. RAMOUS

CONTENTS

APPLIED KINEMATICS

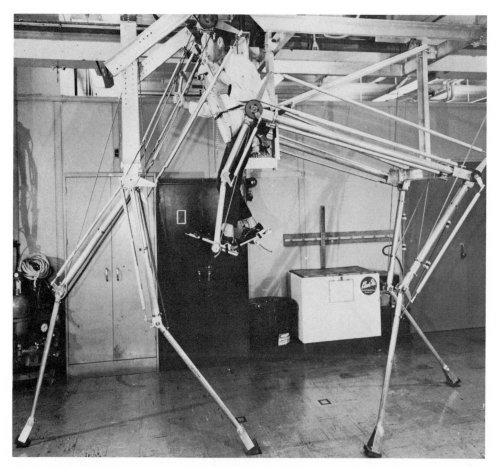

The walking truck is an experimental unit to prove the feasibility of the "mechanism-cybernetic" control concept. The operator literally becomes part of the machine. The operator experiences forces proportional to those imposed upon the machine. (*Courtesy of U. S. Army Tank Automotive Center and General Electric Company.*)

Chapter **1**

MECHANISM MOTION

1.1 Introduction

Kinematics is the science of motion without regard to the forces producing that motion. This book will be concerned with the motion of machine members and mechanisms.

A mechanism is the combination of two or more machine members which function together to perform a specific motion. The term *machine* has a broader meaning than *mechanism*. A machine is a device, usually consisting of several mechanisms, which converts energy into work. A good example of a machine is the internal combustion engine, which converts chemical energy (gasoline) into rotary motion. Another example of a machine is the automatic screw machine shown in Fig. 1.1. Here rotary power is converted into useful machining operations, such as turning, centering, drilling, reaming, and threading. It can be seen that numerous mechanisms are required to accomplish these tasks: linkages, cams, chains, gears, belts, etc.

3

Fig. 1.1. *Automatic screw machine. This model was first built by Automat at the request of the Watchmaker's School in Solothurn, Switzerland, for training their young engineers.* (Courtesy of Automat Precision Engineering Ltd.)

1.2 Link Definition

A mechanism consists of a combination of machine members called links. A link can be a gear, a lever, a cam, or, in a general sense, a belt.

Figure 1.2(a) shows a mechanism link. A line representation for this link is illustrated in Fig. 1.2(b). Points A, B, and C indicate turning or pivot locations. The link is specified as ABC, AB, or BA. Sometimes it is convenient to call link ABC simply $\textcircled{E}$ or body $\textcircled{E}$. The circle around the letter identifies a link, and in this case $\textcircled{E}$ can stand for equalizer bar. Another notation quite frequently used is $\overline{AB}$; the line over the letters is used to emphasize a straight-line distance. This distance need not be on one link but may involve several links.

To simplify kinematic analysis, the assumption is made that a link is non-deformable. This is called the rigid-body assumption. The justification for this assumption is that a load-carrying link is usually only slightly distorted. There are kinematic applications where the link deformation must be considered. For

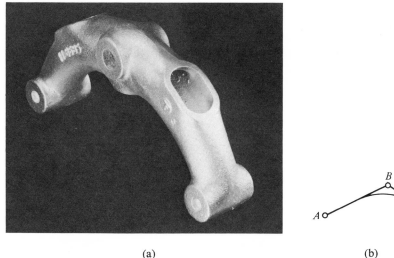

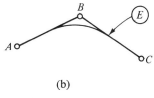

(a)   (b)

Fig. 1.2. (a) Mechanism link (equalizer bar). (b) Kinematic representation of a link. (Courtesy of Malleable Founders Society.)

these cases the rigid-body assumption may serve as the first approximation.

Figure 1.3 shows a link with forces F_1 and F_2 tending to decrease length L. For kinematic analysis, L is considered constant. Be aware that a strength analysis of machine member AB will show a decrease in L.

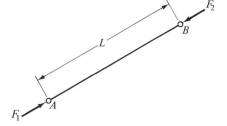

Fig. 1.3. Link AB acted on by forces F_1 and F_2.

1.3 Common Linkages

Figure 1.4 shows a common mechanism called a four-bar linkage. Points A and D are connected to supports, which are in turn connected to the ground.

Fig. 1.4. Four-bar linkage, consisting of links AB, BC, CD, and DA.

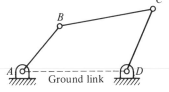

Ground link

It is convenient to think of the ground link AD as stationary. It should be emphasized that the actual situation may show the "ground link moving."

For example, Fig. 1.5(a) shows a four-bar linkage $ABCD$ attached to the floor of an elevator. While the elevator moves vertically upward, point B moves to B'. The basement floor is assumed to be ground. Displacement of B with respect to ground is shown as S_1.

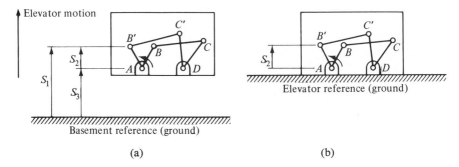

Basement reference (ground)

(a)

Elevator reference (ground)

(b)

Fig. 1.5. (a) *Moving elevator and four-bar linkage. Basement assumed as stationary reference.* (b) *Link* AD *or elevator floor assumed as stationary reference.*

If we are interested in the four-bar linkage displacement relative to the elevator, Fig. 1.5(b) would be more appropriate. In this figure the elevator floor is considered stationary, or as the ground link. Displacement of B with respect to ground is shown as S_2. Figures 1.5(a) and (b) can be related by the following statement:

$$S_1 \quad = \quad S_2 \quad + \quad S_3$$

| Displacement with respect to basement | Displacement with respect to elevator floor | Displacement of elevator floor with respect to basement |

The foregoing example showed that a link (or body) may, for convenience, be considered as stationary. This stationary link (or body) is commonly referred to as ground or earth. The physical concept of being stationary is difficult to comprehend when one considers that no one has experienced complete stillness—the earth we stand on is not stationary but moves in the universe.

Another type of mechanism commonly encountered is the slider crank, Fig. 1.6(a). Crank AB rotates about A, and A is fixed to ground $\textcircled{G}$. Link $\textcircled{S}$ or slider $\textcircled{S}$ moves horizontally between the stationary guides. Figure 1.6(b) shows the slider crank as usually drawn. One of the ground symbols, at the slider, is omitted. It is understood that the slider must stay in contact with the bottom guide at all times. In this particular illustration the slider cannot move vertically, only horizontally.

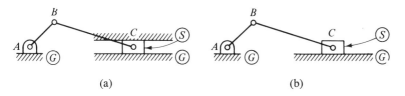

Fig. 1.6. (a) Slider crank. (b) Slider crank (simplified).

1.4 Scale Factors

It is suggested that the reader use a drafting scale similar to the one shown in Fig. 1.7. One side is divided into fractional form, with the smallest marked divi-

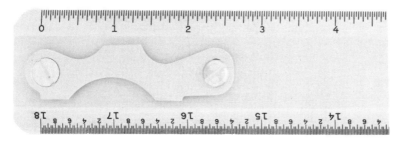

Fig. 1.7. End section of drafting machine scale. (Courtesy of Keuffel and Esser Co.)

sion being $\frac{1}{32}$ in. The smallest marked division on the other scale is .02 in., where .01 in. can be estimated. For mechanisms which are drawn smaller or larger than full size it may be easier to use an architect's scale or a mechanical engineer's scale. By using these scales, one avoids the tedious task of mentally converting machine dimensions into drawing distances.

An acceleration solution may require as many as three layouts (mechanism, velocity, and acceleration). To simplify matters, a scale factor will be assigned to each layout. A scale factor is simply a conversion factor. For example, a scale factor for a mechanism may be 2 in./1 in., commonly referred to as half-size. That is, a machine member 2 in. long is represented on paper as 1 in. The mechanism scale factor is symbolized as K_M; thus $K_M = 2$ in./1 in.

A list of scale factors commonly used are

K_M = mechanism scale factor
K_V = velocity scale factor
K_A = acceleration scale factor

To illustrate how the scale factor is used, suppose a point A has a velocity of 16 mph and it is desirable to place this information graphically on paper. What distance will be used to represent 16 mph? Assume $K_V = 5$ mph/1 in.

$$\text{Drawing distance} = 16 \text{ mph} \left(\frac{1}{K_V}\right) = 16 \text{ mph} \left(\frac{1 \text{ in.}}{5 \text{ mph}}\right) = 3.2 \text{ in.}$$

Let us take another example. A velocity layout shows that a point X has a velocity of 3.65 in. If $K_V = 8$ ft/sec/1 in., what is the velocity in feet per second for point X?

$$\text{Velocity of } X = 3.65 \text{ in. } (K_V) = 3.65 \text{ in. } \left(\frac{8 \text{ ft/sec}}{1 \text{ in.}}\right) = 29.20 \text{ ft/sec}$$

1.5 Slider Cranks

The basic slider crank finds application in many areas, including pumps, internal combustion engines, quick-return mechanisms, and clamping devices. Figure 1.8 shows several positions of a slider crank as AB rotates cw(clockwise)

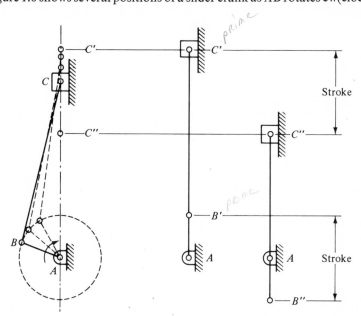

Fig. 1.8. *Extreme positions for slider crank.*

about A. If crank AB rotates through 360 deg, the center of the slider, point C, moves back and forth between two extreme positions. These positions are labeled C' and C''. Distance $C'C''$ is referred to as the stroke. To avoid clashing between links AB and BC, it is necessary to offset these links so that they move in different planes. It can be observed in Fig. 1.8 that the stroke length equals $\overline{B'B''}$, and since $2(\overline{AB}) = \overline{B'B''}$, $\overline{AB} = \frac{1}{2}(\text{stroke})$. This relationship does not hold for

Fig. 1.9. *This clamp is basically a slider crank mechanism.* (Courtesy of De-Sta-Co Corp.)

an offset slider crank. For an offset slider crank the line of motion for point C does not pass through pivot point A (Fig. 1.10). Determine the crank length of a clamp, similar to that shown in Fig. 1.9, which has a stroke of 6 in. Using the relationship $AB = \frac{1}{2}(\text{stroke})$, crank $AB = \frac{1}{2}(6) = 3$ in.

EXAMPLE 1.1

Given: Figure 1.10(a) shows a simplified sketch of a pump mechanism. $AB = \frac{1}{4}$ in., $BC = 2$ in., and eccentricity $= \frac{3}{4}$ in.

Determine: The stroke of the piston (graphically).

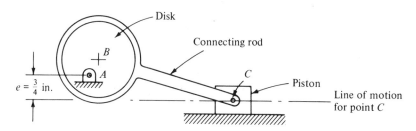

Fig. 1.10. (a) Pump mechanism.

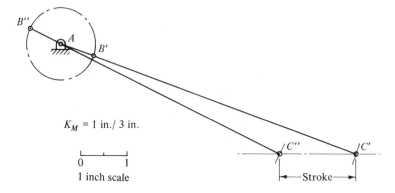

K_M = 1 in./ 3 in.

1 inch scale

Stroke

Fig. 1.10. (b) Extreme positions for offset slider crank.

Solution. This mechanism is equivalent to an offset slider crank. As the disk rotates about point A, the ring portion of the connecting rod slides around the disk. This action produces a reciprocating motion for point C.

STEP 1. To lay out the mechanism, a scale factor must be assumed. Let $K_M = 1$ in./3 in. (three times size); thus on the drafting layout, $AB = \frac{3}{4}$ in., $BC = 6$ in., and eccentricity $= 2\frac{1}{4}$ in. Note that the completed solution is shown in Fig. 1.10(b).

STEP 2. The extreme right position for point C occurs when BC and AB form a straight line. $\overline{AC'} = BC + AB = 2 + \frac{1}{4} = 2\frac{1}{4}$ in. Note that the scale factor was not used here, these are actual dimensions. Using radius $\overline{AC'}$, swing an arc from point A which intersects the line of motion for point C. The intersection establishes point C'.

STEP 3. The extreme left position occurs when BC overlaps AB. $\overline{AC''} = BC - AB = 2 - \frac{1}{4} = 1\frac{3}{4}$ in. To locate C'', swing radius AC'' from A. The intersection with the line of motion establishes C''.

STEP 4. From the layout, measure off distance $C'C''$, $C'C'' = 1.63$ in. The full size length is 1.63 in. (1 in./3in.) $= .54$ in. Stroke is .54 in.

1.6 Examples of Four-Bar Linkage Motion

Examples 1.2 to 1.4 are presented as an introduction to Sect. 1.7, "Four-Bar Linkages."

EXAMPLE 1.2

Given: Figure 1.11(a), four-bar linkage.

Determine: Extreme right position for link DC.

Solution. Point C must move in an arc about D because D is fixed. The path of point C is also established by links AB and BC. The extreme right position of C occurs when AB and BC form a straight line whose distance is $AB + BC$.

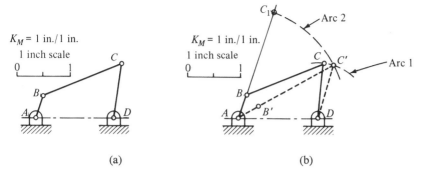

(a)   (b)

Fig. 1.11. (a) Four-bar linkage. (b) Extreme right position of link DC.

STEP 1. Refer to Fig. 1.11(b). Using point D as center, construct arc 1, which passes through point C.

STEP 2. Extend line AB. From point B mark off distance BC onto the extended line; call this point C_1.

STEP 3. Using point A as a center, construct arc 2, which passes through point C_1 and intersects arc 1. The intersection of arc 1 and arc 2 establishes point C', which is the extreme right position of point C.

Example 1.2 can be solved mathematically in the following fashion.

EXAMPLE 1.3

Given: Four-bar linkage, Fig. 1.12, $AB = .44$ in., $BC = 1.62$ in., $CD = 1.00$ in., and $DA = 1.50$ in.

Fig. 1.12. Extreme right position of link DC.

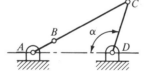

Determine: The extreme right position for link DC (calculate α).

Solution. Use the cosine law:

$$(AC)^2 = (AD)^2 + (CD)^2 - 2(AD)(CD) \cos \alpha \qquad (1)$$

where $AC = AB + BC = .44 + 1.62 = 2.06$ in. Substitute into Eq. (1) and solve for $\cos \alpha$:

$$(2.06)^2 = (1.50)^2 + (1.00)^2 - 2(1.50)(1.00) \cos \alpha$$

$$\cos \alpha = -.33, \qquad \alpha = 109.3°$$

When $\cos \alpha$ is negative, α is larger than 90 deg.

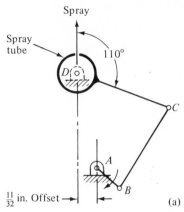

(a)

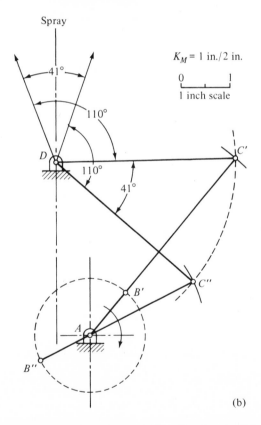

(b)

Fig. 1.13. (a) Oscillating lawn sprinkler utilizes four-bar linkage ABCD. (b) When dial is in "center" position, the extreme positions for link DC are DC' and DC''. (Courtesy of L. R. Nelson Manufacturing Co., Inc.)

12

EXAMPLE 1.4

Given: Figure 1.13(a) shows a photograph and a simplified drawing of an oscillating lawn sprinkler. The dimensions are $AD = 1\frac{3}{4}$ in., $AB = \frac{9}{16}$ in., $CB = 1\frac{3}{4}$ in., and offset $= \frac{11}{32}$ in. When the dial is locked in "center" position, the link length $DC = 1\frac{27}{32}$ in., and the angle between the spray and DC is 110 deg. Crank AB rotates continuously about A.

Determine: Oscillating angle through which the spray moves when the dial is in "center" position.

Solution. This mechanism can be recognized as four-bar linkage $ABCD$, where the spray tube is part of link DC. The 110-deg angle between DC and the spray is maintained throughout the cycle.

It is interesting to note that the power to drive crank AB is obtained from the same water which wets down the lawn. The water entering the sprinkler drives a small impeller, which in turn drives a double set of worm and worm gears. The worm and worm gears reduce speed and at the same time increase torque on crank AB.

STEP 1. The construction is shown in Fig. 1.13(b). Assume $K_M = 1$ in./2 in. (double size). Lay out points A and D.

STEP 2. Using radius DC, swing an arc from D. The path of motion for point C is on this arc.

STEP 3. The highest position for DC occurs when CB and AB form a straight line. $AC' = CB + AB = 1\frac{3}{4} + \frac{9}{16} = 2\frac{5}{16}$ in. Using radius AC', swing an arc from point A which intersects the arc of motion for point C. This establishes DC'.

STEP 4. The lowest position for DC occurs when CB and AB overlap. $AC'' = CB - AB = 1\frac{3}{4} - \frac{9}{16} = 1\frac{3}{16}$. Swing radius AC'' from A; the intersection with the arc of motion establishes point C''. The angle between DC' and DC'' is found to be 41 deg.

STEP 5. Since the spray tube is part of link DC, the spray tube and spray oscillate through 41 deg. The spray leaves the tube 110 deg from DC, as shown in Fig. 1.13(b). Note that the 41 deg that the spray moves through is approximately centered with the vertical center line.

1.7 Four-Bar Linkages

Four-bar linkages are used in windshield wiper mechanisms, typewriters, mechanical calculators, clamping devices, and many other mechanisms.

Figure 1.14 shows three different four-bar linkages. The linkage shown in Fig. 1.14(a) is called a crank-rocker or crank and rocker mechanism. Crank AB rotates through 360 deg, while the rocker CD oscillates twice through angle θ. The crank and rocker are connected by link BC, which is called the coupler or connecting rod. Crank AB is the shortest link in a crank-rocker mechanism. Considering AB to rotate at a constant rpm ccw (revolutions per minute counterclockwise), it takes less time for link CD to move from position $C'D$ to $C''D$ (fast) than from $C''D$ to $C'D$ (slow). This can be simply demonstrated by assuming that AB rotates at 1 deg/sec; thus a crank rotation of 360 deg requires 360 sec.

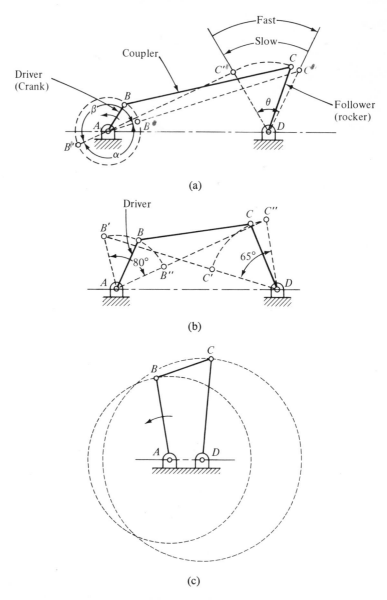

Fig. 1.14. Four-bar linkages. (a) Crank-rocker mechanism. (b) Double-rocker mechanism. (c) Drag-link mechanism.

Since the crank angle $\alpha = 171$ deg, it takes 171 sec for CD to move from $C'D$ to $C''D$. $C''D$ to $C'D$ requires 189 sec, $\beta = 189$ deg. A crank-rocker mechanism can be applied to quick-return applications where a slow advance and rapid return is required.

A variation of Fig. 1.14(a) is to have link CD the driver and AB the follower. This mechanism has limited use because it is difficult to drive.

If links AB and CD both oscillate, the linkage is called a double-rocker mechanism. Figure 1.14(b) shows the extreme positions for a double-rocker mechanism with link AB driving. It is understood that an oscillating motor or some other device which is not shown is driving link AB. This type of mechanism can be used to increase or decrease the angle of oscillation. For example, Fig. 1.14(b) shows a decrease in the angle of oscillation, the input angle is 80 deg and the output angle is 65 deg. A crank-rocker mechanism, Fig. 1.14(a), can be operated as a double-rocker mechanism.

When links AB and CD rotate continuously through 360 deg, as in Fig. 1.14(c), the linkage is called a drag-link or double-crank mechanism. If the driver crank rotates at a constant rpm, the follower rotates at a variable speed. In a drag-link mechanism the ground link AD is the shortest link. This linkage can be used as a quick-return mechanism.

The parallel four-bar linkage shown in Fig. 1.15 does not fit into the categories previously discussed. Here $AB = CD$ and $AD = BC$. It is possible to design this mechanism so that links AB and CD rotate together at the same speed.

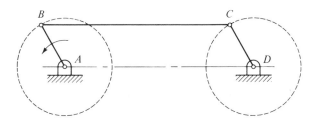

Fig. 1.15. *Parallel four-bar linkage.*

1.8 Dead Positions

A mechanism is said to be in a dead position or dead phase when the driver theoretically cannot move the mechanism.

Figure 1.16(a) shows a simplified sketch of a hand-driven slider crank. Motion is applied to the handle link, and rotary motion is expected from crank AB. The handle link is slotted to allow point C to move horizontally. When the slider crank is in position $AB'C'$, Fig. 1.16(b), the force at the slider cannot drive the crank because the connecting rod and crank are in line. Assuming perfect alignment, all that the applied force does is push on ground connection A. This is a dead position for the slider crank and is commonly referred to as top dead center (TDC). Another dead position occurs at $AB''C''$, Fig. 1.16(b). This is called bottom dead center (BDC).

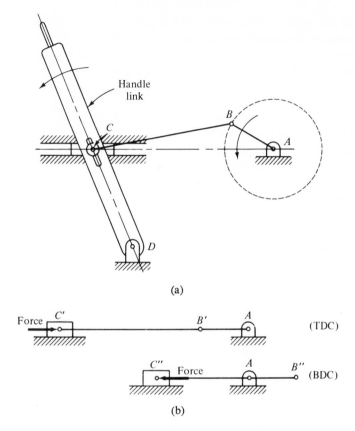

(a)

(b)

Fig. 1.16. *Dead positions for slider crank mechanism, slider driving.*

This mechanism can be made to operate by attaching a flywheel to crank AB. This should carry the linkage past the dead positions. The handle link cannot start the crank or flywheel moving from a dead position; therefore, one should avoid stopping the mechanism in a dead position.

If in Fig. 1.16(a) crank AB had been the driver, the mechanism would not have dead positions. Crank AB can easily move through positions $AB'C'$ and $AB''C''$. Generally before dead positions can be determined for a mechanism, one must know which link is driving.

Figure 1.17 shows a four-bar linkage in a dead position where AB is driving. Link AB cannot move to the left without breaking a link or pin connection.

Fig. 1.17. *Dead position for a four-bar linkage,* AB *driving.*

Also, link AB cannot move to the right because BCD forms a straight line. Actually, owing to a slight misalignment along BCD, it is possible that the linkage will buckle at C. It cannot be readily predicted in which direction point C will buckle. To avoid this unpredictable situation, stop the linkage with a mechanical stop before the dead position is reached. It should be noted that this is not a dead position for the mechanism when CD is driving.

Many locking mechanisms are designed to stop at or near the dead position of a linkage. Locking linkages are commonly used to hold open the tops of attaché cases, luggage, and phonographs.

The characteristics of a dead position are such that a high clamping force is usually obtainable. Linkages utilizing this feature are called toggle mechanisms; Fig. 1.9 shows such a clamping device.

1.9 Kinematic Synthesis
of a Four-Bar Linkage

So far we have been concerned with the analysis of linkages. That is, given the linkage dimensions, determine the motion of the linkage. A far more difficult problem and one which is usually faced by the designer is, given the motion, determine the linkage dimensions. Since we consider only kinematic properties of the mechanism, this aspect of design is called *kinematic synthesis*. Synthesis is the opposite of analysis.

The selection of a satisfactory mechanism to accomplish the required motion is sometimes complicated by the fact that there are many mechanisms which can produce the same motion. Take, for example, straight-line motion; straight-line motion can be produced with a slider crank, four-bar linkage, gear and rack, cam and follower, planetary mechanism, plus many other mechanisms. Further, nonmechanism methods are available, such as a fluid-operated piston or a linear solenoid. If the known mechanisms cannot fulfill the required motion, the designer must either modify an existing mechanism or invent a mechanism.

A typical kinematic design problem is shown in Fig. 1.18(a). Here it is required that link CD move from $C'D$ to $C''D$ and back again to $C'D$. This motion is to be synchronized to one revolution of the input shaft at A. Quantities AD, CD, ϕ, and θ are known. A four-bar linkage $ABCD$, crank-rocker type, will be used to accomplish this motion. Thus we require lengths for AB and BC. Before the solution is discussed, let us look at a crank-rocker mechanism, Fig. 1.18(b). When CD is in the extreme right position, links AB and BC are outstretched. For the extreme left position of CD, links AB and BC are overlapping. A careful comparison of these two extreme positions shows that $AB''C''$ is longer than AC' by twice AB.

Stated mathematically, $AB''C'' = AC' + 2(AB)$. This is the basis for the graphical solution that follows.

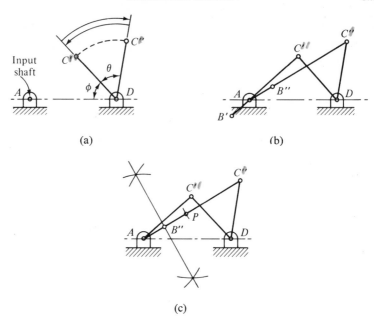

(a)   (b)

(c)

Fig. 1.18. (a) Given information. (b) Extreme positions for
crank-rocker mechanism. (c) Construction of link dimensions
AB and BC for a crank-rocker mechanism.

STEP 1. The solution is shown in Fig. 1.18(c). Mark off AC' onto AC'' from point
C''. This establishes point P.

STEP 2. Divide AP into two equal parts (perpendicular bisector technique). This
locates position B''.

STEP 3. Finally, scale off AB and BC along line $AB''C''$.

Example 1.5 shows the mathematical solution to this type of problem.

EXAMPLE 1.5

Given: Link CD, Fig. 1.19, is to oscillate through $\theta = 55$ deg while a shaft at A
rotates through 360 deg. $AD = 1.5$ in., $CD = 1.0$ in., and $\phi = 45$ deg.

Determine: The link lengths AB and BC for a crank-rocker mechanism $ABCD$.

Solution. Apply the cosine law to linkage triangle $AB''C''D$, Fig. 1.19:

$$(AB''C'')^2 = (C''D)^2 + (AD)^2 - 2(C''D)(AD) \cos (\phi + \theta) \tag{1}$$

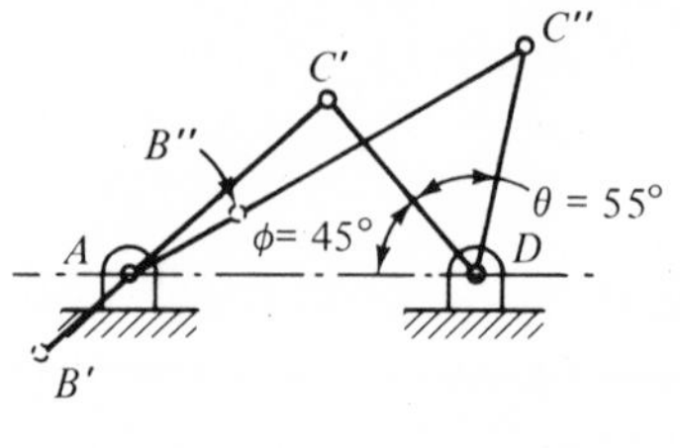

Fig. 1.19. Point B was arbitrarily chosen to depict typical
extreme positions for crank-
rocker ABCD.

where $C''D = CD = 1.0$ in., $AD = 1.5$ in., and $\phi + \theta = 45$ deg $+ 55$ deg $= 100$ deg. Substitute numerical values into Eq. (1):

$$(AB''C'')^2 = (1.0)^2 + (1.5)^2 - 2(1.0)(1.5) \cos 100°$$
$$AB''C'' = 1.940 \text{ in.,} \quad \text{or}$$
$$AB''C'' = AB + BC = 1.940 \text{ in.} \tag{1a}$$

Apply the cosine law to linkage triangle $AC'D$:

$$(AC')^2 = (C'D)^2 + (AD)^2 - 2(C'D)(AD) \cos \phi \tag{2}$$
$$(AC')^2 = (1.0)^2 + (1.5)^2 - 2(1.0)(1.5) \cos 45°$$
$$AC' = 1.063 \text{ in.,} \quad \text{or}$$
$$AC' = BC - AB = 1.063 \text{ in.} \tag{2a}$$

Add Eq. (2a) to Eq. (1a); this eliminates AB:

$$BC + AB = 1.940$$
$$\underline{BC - AB = 1.063}$$
$$2(BC) = 3.003$$
$$BC = 1.502 \text{ in.}$$
$$\text{and} \quad AB = .438 \text{ in.}$$

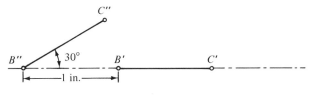

Fig. 1.20. (a)

EXAMPLE 1.6

Given: Figure 1.20(a); link BC is to move from $B'C'$ to $B''C''$. Link $BC = 1$ in.

Determine: A four-bar linkage $ABCD$ to accomplish this motion. Lengths AD, AB, and CD are required.

Solution. The completed solution is shown in Fig. 1.20(b).

STEP 1. Assume a mechanism scale; let $K_M = 1$ in./2 in. Draw perpendicular bisectors for $B''B'$ and $C''C'$.

STEP 2. Arbitrarily select points A and D on perpendicular bisector lines. Draw in four-bar linkage $ABCD$.

STEP 3. Scale off link dimensions. $AD = .50$ in., $AB = .88$ in., and $CD = 1.22$ in.

Had points A and D both been chosen at P, the intersection of the perpendicular

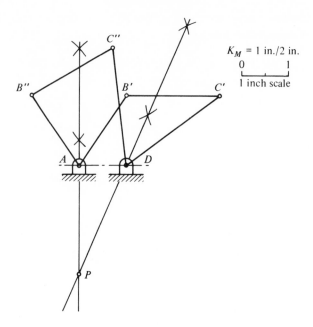

$K_M = 1$ in./2 in.

1 inch scale

Fig. 1.20. (b) Design of a four-bar linkage.

bisectors, links PB, BC, and CP would form one rigid link. With point P as the pivot point, link PBC can easily rotate from position $PB'C'$ to $PB''C''$.

1.10 Coupler Curves

Another type of kinematic synthesizing involves the motion of a point on the coupler link of a four-bar linkage. The curve traced by the coupler point is characteristically called a coupler curve. The coupler point can be used directly as an output or may be used to drive other links.

Figure 1.21 shows a model of a crank-rocker mechanism. A lead pencil traces the coupler curve on a piece of paper. Different curves can be obtained by altering the lengths of the linkage and the relative position of the coupler point. A mechanism similar to the one shown in Fig. 1.21 is used in a dough-mixing machine. The output motion resembles that of a baker mixing dough.

Figure 1.22 illustrates three possible design arrangements which utilize the same coupler curve. In Fig. 1.22(a) only a portion of the coupler curve is desirable—that from P' to P''. One method of obtaining this motion is to drive linkage $ABCD$ with a crank-rocker mechanism $EFCD$. Crank EF rotates 360 deg while link CD oscillates through the appropriate angle θ.

In Fig. 1.22(b) an approximate dwell, corresponding to a crank angle of 90 deg, can be obtained by having the coupler point drive a slotted link $\textcircled{L}$. Link $\textcircled{L}$ is pivoted at E. The dwell portion on the coupler curve is from P' to

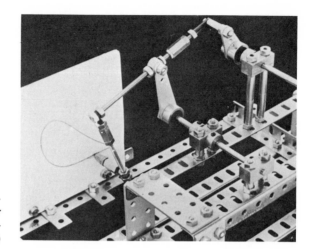

Fig. 1.21. *Coupler curve generated by a crank-rocker mechanism.* (Courtesy of Automat Precision Engineering Ltd.)

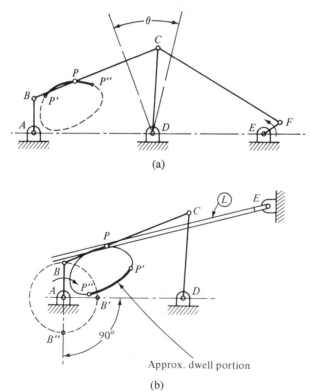

(a)

(b)

Fig. 1.22. (a) *Linkage* ABCD *is driven by crank rocker* EFCD. (b) *Dwell mechanism.*

21

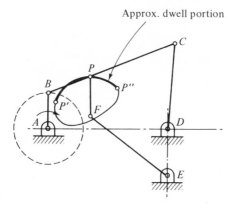

Approx. dwell portion

Fig. 1.22. (c) *Dwell mechanism.*

P''. During the dwell the link $\underline{L}$ practically stops moving while crank AB continues to rotate.

A dwell can be obtained for link FE from the six-bar linkage shown in Fig. 1.22(c). Coupler path P' to P'' is practically a circular arc. Link PF is made equal to the circular arc's radius. As P moves from P' to P'', link PF essentially rotates about point F. Since point F is not moving during this portion of the cycle, FE dwells.

For a useful atlas on coupler curves, the reader can refer to *Analysis of the Four-Bar Linkage* by John A. Hrones and George L. Nelson, published jointly by John Wiley & Sons, Inc., and MIT Press, 1951.

1.11 Eccentric Cams

A circular disk which rotates about an axis which is not the center of the disk is called an eccentric cam or simply an eccentric. Eccentric cams are easily synthesized.

In Fig. 1.23 an eccentric cam is shown driving a flat-face follower. Illustrated are the high and low positions of the follower. The cam or disk radius is symbolized as R, and the distance from the cam center C to the axis of rotation O is the eccentricity e. Distances from O to the follower face are labeled L_1 and L_2. It can be seen that

$$\text{Stroke} = L_1 - L_2 = (R + e) - (R - e) = R + e - R + e$$

Thus

$$\text{Stroke} = 2e$$

This statement is very similar to the stroke-crank relationship for a slider crank.

EXAMPLE 1.7

Given: Figure 1.24(a); a roller follower is to rise and fall $1\frac{1}{4}$ in. At the lower position, the center of the roller is located 3 in. from the axis of rotation. The roller diameter is 1 in.

Determine: An eccentric cam to accomplish this motion.

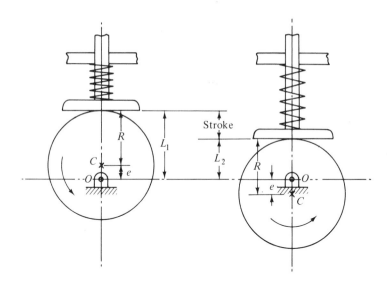

Fig. 1.23. Eccentric cam and flat-face follower.

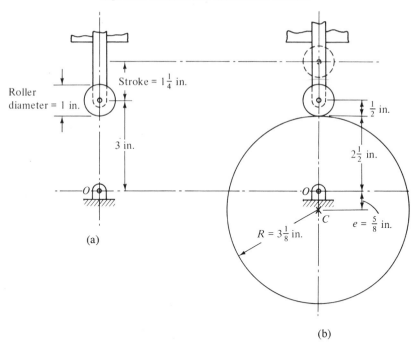

Fig. 1.24. Eccentric cam and roller follower.

Solution. This is a low-speed application where it is assumed that the cam-follower action does not cause excessive binding or wear on the follower stem. The completed solution is shown in Fig. 1.24(b). Note that for the lower position, cam center C is located directly below O. The distance from the axis of rotation to the bottom of the roller is $2\frac{1}{2}$ in. Since the stroke is $1\frac{1}{4}$ in., $e = \frac{1}{2}(1\frac{1}{4}) = \frac{5}{8}$ in. The cam radius is $2\frac{1}{2} + \frac{5}{8} = 3\frac{1}{8}$ in. In summary, the eccentricity is $\frac{5}{8}$ in. and the cam radius is $3\frac{1}{8}$ in.

23

1.12 Kinematic Analysis of Mechanisms

Figure 1.25(a) shows an air-operated arbor press. Kinematic equivalents are shown in Figs. 1.25(b) and (c). Figure 1.25(b) is visually closest to the actual mechanism; however, Fig. 1.25(c) may be preferred because of its simplicity. The rectangle at point C indicates that BC is rigidly joined to CD, forming one link BCD. Referring to Fig. 1.26, point B moves vertically along path 1, and C moves in an arc about point F, called arc 1. DE does not constitute a rigid link because the distance between D and E changes.

Let us say it is required to determine the piston stroke which corresponds to a ram stroke of B to B'. Figure 1.26 is used to show the displacements.

STEP 1. From point B', mark off distance BC onto arc 1. This locates C'.
STEP 2. Draw a line through B' and C' which extends beyond C'.
STEP 3. From point C', mark off distance CD onto extended line $B'C'$. This locates D'.
STEP 4. From point E, mark off distance ED onto ED'. This locates D_1. The piston stroke is distance D_1D'.

A cylinder capable of delivering a longer stroke would not be detrimental. In fact it would be preferred because the stop nut [Fig. 1.25(a)] can then be adjusted, if desirable, to increase the ram stroke. One should not lose sight of the fact that D physically represents a point on the piston rod and not a point on the piston.

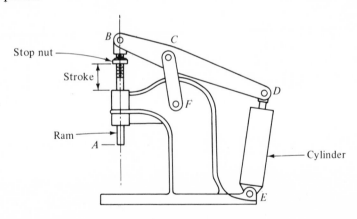

Fig. 1.25. (a) Arbor press.

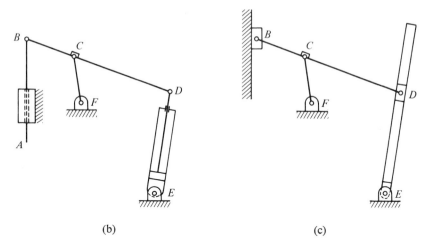

(b) (c)

Fig. 1.25. (b) *Kinematic skeleton of arbor press.* (c) *Kinematic skeleton of arbor press.*

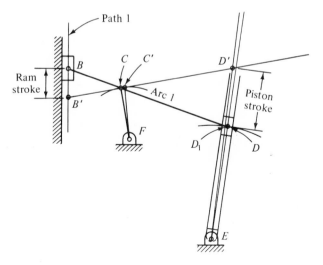

Fig. 1.26. *Arbor press displacements.*

Example 1.8

Given: Figure 1.27(a) shows the open position of a bottom-dump door mechanism used on railroad ore cars. The mechanism is driven by a pneumatic cylinder. The photograph shows both door mechanisms in the closed position, with the pistons extended.

Determine: The closed position for the door; use the piston stroke shown in Fig. 1.27(a).

Solution. The points A, C, and F, attached to the car, are assumed to be ground points. The completed solution is shown in Fig. 1.27(b).

Step 1. Draw a line AB extending beyond B. At B, place the length of the piston stroke on extended line AB. This locates B_1.

Step 2. Using point A as a center of rotation, swing an arc through B_1. Label this arc arc 1.

Step 3. Using C as the center of rotation, swing an arc through B which intersects arc 1. This is the new position for B. Label it B'.

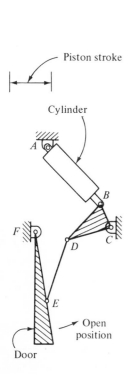

Fig. 1.27. (a) (Courtesy of Baldwin-Lima-Hamilton Corporation.)

STEP 4. Since BCD is one solid link, D is a fixed distance from B and C. Swing an arc of radius BD from B'. Also swing an arc of radius DC from C. The intersection of these arcs locates D'.

STEP 5. Swing an arc from D' of radius ED. Swing an arc from F of radius EF. The intersection of these arcs locates E'. This completes the construction.

For simplicity, the cylinder-piston rod is represented in Fig. 1.27(b) by lines AB and AB'.

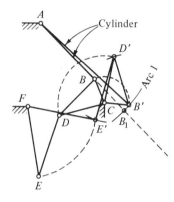

Fig. 1.27. (b)

PROBLEMS

1.1 A slider crank used in an internal combustion engine has a crank length of .875 in. and a connecting rod length of 3.125 in. What is the piston stroke?

1.2 In Fig. P1.2 a slider moves back and forth between C' and C'' as AB continuously rotates. Distance $C''C' = 4$ in. and $AC'' = 8$ in. Graphically determine the lengths for the crank and connecting rod.

Fig. P1.2.

1.3 Do Prob. 1.2 mathematically. *Hint:* Use Eqs. (1) and (2):

(1) $AB' + B'C' = 12$

(2) $B''C'' - AB'' = 8$

1.4 In the position shown in Fig. P1.4, the compression spring exerts a 5-lb force on the slider. Determine the total spring force developed when crank AB moves to position AB'. The spring constant is $10\ \text{lb}/1$ in. $AB = \frac{7}{8}$ in., $BC = 3\frac{3}{8}$ in., and the slider length is $\frac{1}{2}$ in.

Fig. P1.4.

1.5 For the angular position shown in Fig. P1.5 (15 deg), determine the stroke for the power hack saw. Length $AB = 3$ in. and $BC = 8\frac{1}{2}$ in.

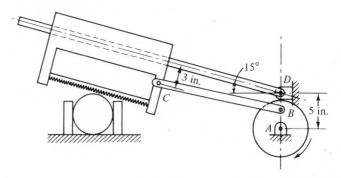

Fig. P1.5.

1.6 Determine the stroke for the slider shown in Fig. P1.6. Crank AB rotates 360 deg about point A. Rack link CD always stays in contact with the spur gear. $AB = 1$ in., $BC = 3$ in., $CD = 4\frac{1}{2}$ in., and the gear diameter is 1 in.

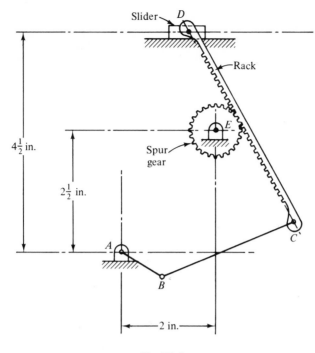

Fig. P1.6.

1.7 Determine graphically or mathematically the extreme right position for link *AB*, Fig. P1.7. *AD* = 3.0 in., *CD* = 1.5 in., *CB* = 2.0 in., and *BA* = 1.0 in.

1.8 Check the solution to Prob. 1.7 using a four-bar linkage model. Links can be cut from stiff cardboard, plastic, wood, or tin.

1.9 Determine graphically the extreme left position for point *A*, Fig. P1.9. Explain the type of motion *AB* has. *AB* = 4.0 in., *BC* = 2.0 in., *CD* = 4.0 in., and *DA* = 2.0 in.

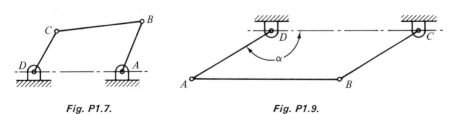

Fig. P1.7.   **Fig. P1.9.**

1.10 Determine mathematically the extreme left position for point *A* in Prob. 1.9. The angle α is required.

1.11 Figure P1.11 shows a simplified drawing of an Aqua-Queen oscillating lawn sprinkler. The spray pattern can be changed by adjusting angle θ and length CD; the adjusting mechanism is not shown. The spray tube is part of link CD; this tube is shown in cross section. A small hydraulic motor drives AB through 360 deg. Points A and D are stationary. For various values of CD and θ, determine the oscillating angle through which the spray moves.

Spray coverage	DA (in.)	AB (in.)	BC (in.)	CD (in.)	θ (deg.)	Oscillating angle (deg.)
Full	$1\frac{5}{8}$	$\frac{5}{8}$	$1\frac{5}{8}$	$\frac{7}{8}$	102	
Left	$1\frac{5}{8}$	$\frac{5}{8}$	$1\frac{5}{8}$	$1\frac{1}{2}$	143	
Right	$1\frac{5}{8}$	$\frac{5}{8}$	$1\frac{5}{8}$	$1\frac{7}{8}$	100	
Narrow	$1\frac{5}{8}$	$\frac{5}{8}$	$1\frac{5}{8}$	$2\frac{1}{8}$	130	

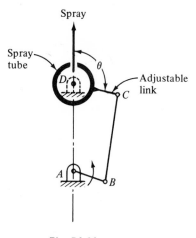

Fig. P1.11.

1.12 For the windshield wiper mechanism shown in Fig. P1.12, determine the angle of oscillation for each blade. An electric motor continuously drives link ED. The dimensions are $AB = FG = ED = 1$ in., $AF = BG = 16$ in., $BC = \frac{7}{8}$ in., and $CD = 7\frac{5}{8}$ in. ABC is one link with a right angle at B. A tension spring is located in the center of link BG. Why is this spring required?

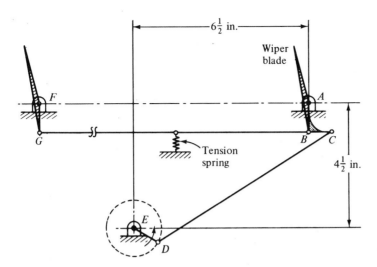

Fig. P1.12.

1.13 A four-bar linkage is used to open and close a rotary valve in a washing machine. Figure P1.13 shows a sectional view of the valve as it would appear in the closed position, with no flow. The gate and link CD are rigidly fastened to the shaft at D. Gate stops are located 90 deg apart. Determine the angle that link AB is required to move through to fully open this valve. $AB = 2\frac{5}{16}$ in., $BC = 3\frac{3}{8}$ in., and $CD = \frac{7}{8}$ in.

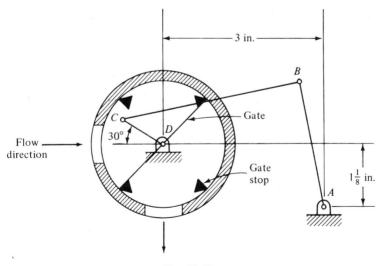

Fig. P1.13.

1.14 Figure P1.14 shows a kinematic skeleton and photograph of a hand-operated wire stripping tool. This tool is used to strip off the insulation on small-diameter electrical wire. Point A (axis) is connected to a movable cutting die which can move $\frac{3}{8}$ in. down before it butts against a stationary cutting die; then the entire top portion of the tool pivots about point C. Determine the final position of the tool. To do this move point D to D''. $AB = BC = 1$ in., $BD = 4\frac{3}{8}$ in., and $CD = 4\frac{1}{4}$ in.

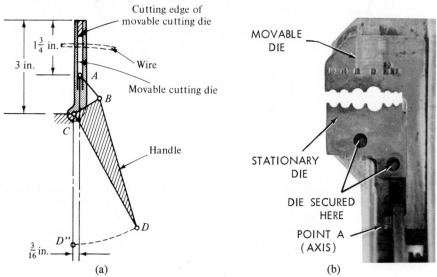

Fig. P1.14. (a) Kinematic skeleton of a wire stripping tool. (b) A side view of (a). A covering plate was removed to show this mechanism. (Courtesy of Ideal Industries, Inc.)

1.15 Figure P1.15 shows a modified slider crank in which point D traces out an elliptical path. BCD is one link. Determine graphically or mathematically AB, BC, and CD. The major axis is to be 6 in. Can D trace the entire elliptical path? Explain. Assume values, if necessary, to solve this problem. *Hint:*

$2(AB + BC + CD) = 6$

$2(CD) = $ minor axis

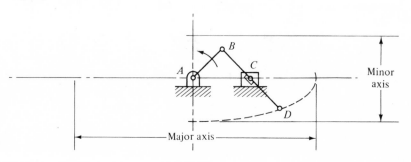

Fig. P1.15.

1.16 In Fig. P1.16, link ABC of a Scotch-yoke mechanism rotates 360 deg ccw. A pin is rigidly attached to ABC at B. Calibrate the angle θ and $\cos \theta$ scales. For example, when pointer at C indicates 150 deg, the pointer at D should read $-.866$. Link $AB = 1$ in.

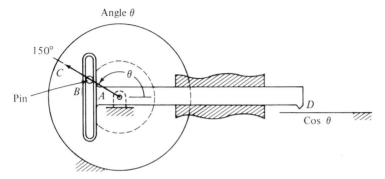

Fig. P1.16.

1.17 In Fig. P1.17, a wheel rolls on the ground without slipping. Determine the path of point P for one revolution of the wheel. The solution has been started in the figure.

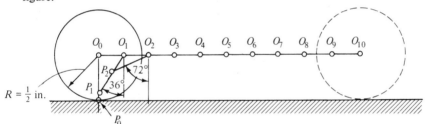

Fig. P1.17.

1.18 Figure P1.18 shows a drag-link mechanism. Plate F is rigidly attached to shaft E and rotates about axis A. Plate G is rigidly fastened to H and rotates about axis D. Link J is pinned to F and G at B and C, respectively. Determine the position of point P for every 30 deg of rotation of shaft H. $AD = \frac{7}{16}$ in., $AB = \frac{1}{2}$ in., $BC = \frac{7}{8}$ in., $CD = \frac{13}{16}$ in., $CP = \frac{7}{8}$ in., and $BP = \frac{7}{8}$ in.

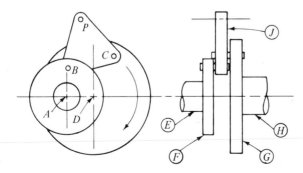

Fig. P1.18.

1.19 In Fig. P1.19, crank AB rotates at a constant 100 rpm while CD oscillates. What direction of rotation is required for AB so that CD moves faster (less time) to the right. Also, determine the total time it takes for CD to move from the extreme left to the extreme right position. $AB = 1\frac{1}{4}$ in., $BC = 3\frac{1}{4}$ in., $CD = 2$ in., and $DA = 3$ in.

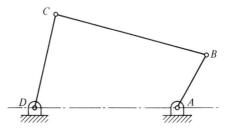

Fig. P1.19.

1.20 In Fig. P1.20, two identical slider cranks are connected to a slider at point C. Assuming that AB is the driver and can rotate continuously, what are the dead positions for the mechanism? What link should we attach the flywheel to? If the slider were driving the entire mechanism, where would it be best to attach the flywheel? $AB = ED = \frac{1}{2}$ in. and $BC = CD = 3\frac{1}{2}$ in.

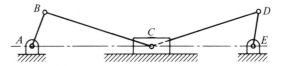

Fig. P1.20.

1.21 Determine the dead positions for a four-bar linkage $ABCD$ when AB is driving. Also, determine the dead positions when CD is driving. Ground link is AD. $AB = 1.0$ in., $BC = 2.0$ in., $CD = 1.5$ in., and $DA = 3.0$ in.

1.22 Determine the extreme positions for the mechanism shown in Fig. P1.22. Consider the wheel as the driver. Are there any dead positions? $AB = 3$ in., $BC = 3\frac{1}{2}$ in., $CD = 1\frac{1}{2}$ in., $DA = 3$ in., $CE = 1$ in., $EF = 2$ in., and the wheel diameter $= 4$ in. BEC is one link.

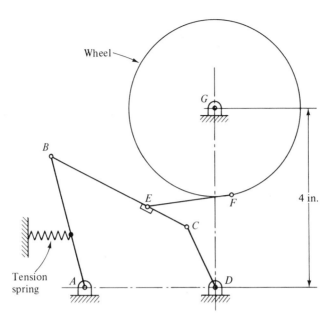

Fig. P1.22.

1.23 Figure P1.23 shows a four-bar linkage mounted on a platform. The platform moves vertically at a constant velocity of $\frac{1}{2}$ in./sec. Crank AB rotates at a constant angular velocity of $\frac{1}{12}$ rev/sec. The starting position for the platform and linkage is shown in the figure. Determine the vertical position of point C with respect to the ground $\textcircled{G}$ after 7 sec. $AB = \frac{3}{4}$ in., $BC = 3$ in., $CD = 1\frac{1}{4}$ in., and $DA = 3$ in.

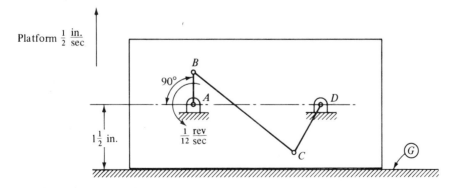

Fig. P1.23.

1.24 In Fig. P1.24, four-bar linkage $ABCD$ carries the four-bar linkage $BFEC$. A motor, secured to the ground, drives link AB ccw. An identical motor is secured to BC and drives CE cw. Assume AB and CE rotate at the same rpm. Determine five positions for point F with respect to ground. Use 20-deg increments for AB and CE. $AB = \frac{1}{2}$ in., $BC = 2\frac{1}{8}$ in., $CD = \frac{3}{4}$ in., $DA = 2$ in., $CE = \frac{1}{2}$ in., $EF = 2\frac{1}{8}$ in., and $FB = \frac{3}{4}$ in.

1.25 Graphically design a crank-rocker mechanism $ABCD$ to oscillate link CD through an angle of 60 deg; see Fig. P1.25. $AD = 2.0$ in. and $CD = 1.5$ in.

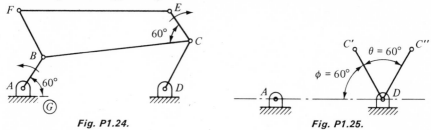

Fig. P1.24.   Fig. P1.25.

1.26 Solve Prob. 1.25 mathematically.

1.27 It is required to have a pointer oscillate through an angle of 60 deg, as shown in Fig. P1.27. This motion is to be accomplished with a crank-rocker mechanism $ABCD$. Determine all dimensions. $AD = 3.0$ in. and the disk diameter $= 2$ in.

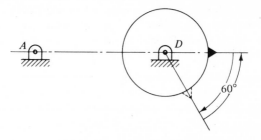

Fig. P1.27.

1.28 It has been decided that a crank-rocker mechanism $ABCD$ is to index a ratchet (Fig. P1.28). The ratchet is 1 in. in diameter and has 12 teeth. The ratchet is to advance one tooth for each revolution of the crank AB. $CD = 1.5$ in. and $AD = 3.5$ in. Determine lengths for AB and BC.

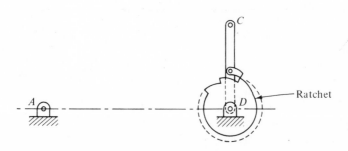

Fig. P1.28.

1.29 In Fig. P1.29, AB of a crank-rocker mechanism is to oscillate through an angle of 70 deg while crank DC rotates 360 deg. Follower link $AB = \frac{7}{8}$ in. Graphically determine CD and CB.

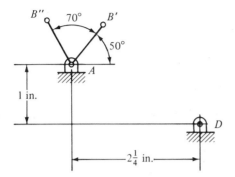

Fig. P1.29.

1.30 Solve Prob. 1.29 mathematically.

1.31 Determine BC and CD for a crank-rocker mechanism (Fig. P1.31). CD oscillates through 90 deg while AB rotates 360 deg. $AB = .75$ in. and $AD = 2.5$ in. *Hint:* use a trial-and-error solution.

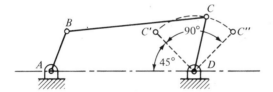

Fig. P1.31.

1.32 In Fig. P1.32, link CB is to move from $C'B'$ to $C''B''$. Link $CB = 1$ in. Determine (a) a four-bar linkage $ABCD$ to accomplish this motion. (b) The same motion is to be accomplished by pivoting one link about a point. Solutions require all link dimensions and pivot locations.

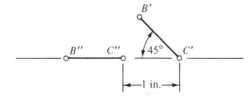

Fig. P1.32.

1.33 Given the same information as in Prob. 1.32 plus the additional requirements that when CB moves from $C'B'$ to $C''B''$, link BA, of a four-bar linkage $ABCD$, moves through an angle of 30 deg ccw and link CD moves through an angle of 30 deg cw. Determine link dimensions and pivot locations for the four-bar linkage.

1.34 Design a four-bar linkage $ABCD$ so that BC can move through three positions $B'C'$, $B''C''$, and $B'''C'''$(Fig. P1.34). Link $BC = 1$ in.

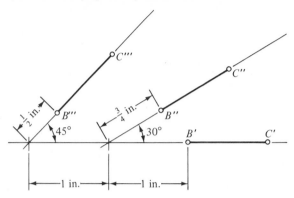

Fig. P1.34.

1.35 Figure P1.35 shows a simplified sketch of an agricultural machine. A road wheel drives crank AB of crank-rocker $ABCD$ by means of a chain drive. The wheel rotates at half the rpm of crank AB. Note that rocker CD is not directly driven by the wheel. The relative sizes are $AB = .50$, $BC = 1.0$, $CD = 2.5$, $DA = 2.5$, and $CBP = 2.0$, and the wheel diameter $= 4.0$. Determine the coupler curve for point P as the wheel rotates through $\frac{1}{2}$ rev. Assume no slip between ground and wheel.

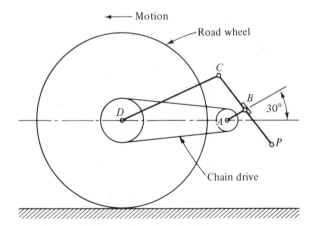

Fig. P1.35.

1.36 A translating flat-face follower is to rise and fall 1 in. The farthest position of the follower face from the shaft center is 2 in. Design an eccentric cam to do this job.

1.37 In Fig. P1.37, a flat-face follower is offset from the shaft center by $\frac{1}{4}$ in. The required stroke is $\frac{3}{4}$ in. The maximum distance from the follower face to the shaft center is 2.5 in. Design an eccentric cam to accomplish this motion.

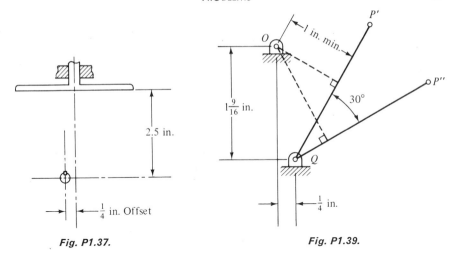

Fig. P1.37.   Fig. P1.39.

1.38 A translating roller follower is to rise and fall 1 in. At the lower position, the center of the roller is located $2\frac{1}{2}$ in. from the axis of rotation. The roller diameter is 1 in. Determine the eccentric cam to accomplish this motion.

1.39 Design an eccentric cam to move QP through a total angle of 30 deg from QP' to QP'' (Fig. P1.39). At the QP' position the perpendicular distance to O is 1 in. Check your design by cutting out a cardboard cam and follower.

1.40 Figure P1.40 shows an oscillating roller follower and eccentric cam. Determine the highest and lowest position for point C. The roller diameter is 1 in.

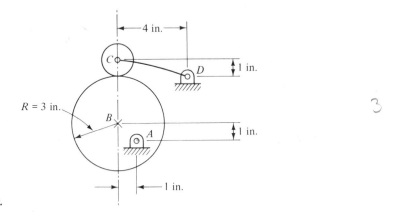

Fig. P1.40.

1.41 Do Prob. 1.40 considering $ABCD$ as a crank-rocker mechanism.

1.42 Figure P1.42 shows the kinematic skeleton of an air-operated arbor press. Determine the piston stroke for a ram stroke of 2 in. $DE = 7.90$ in., $BD = 8.06$ in., and $CD = 6.16$ in.

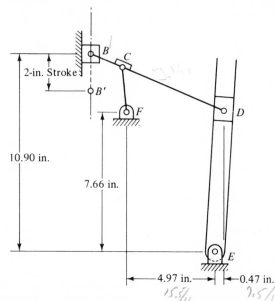

Fig. P1.42.

1.43 Figure P1.43 shows a simplified sketch of a hand-operated mechanism used to emboss metal plates. In the embossing position, two of the three four-bar linkages (*DEFG* and *GHIJ*) come into toggle, producing approximately a 1300-lb force from an 18-lb pull on the handle. The handle is keyed to plate *AB*. Links are pivoted at *A*, *D*, *G*, and *J*. Determine the position of *JI* when the handle is moved 37 deg from the vertical position. Use a 5 to 1 scale.

Fig. P1.43. (Courtesy of Pitney-Bowes, Inc.)

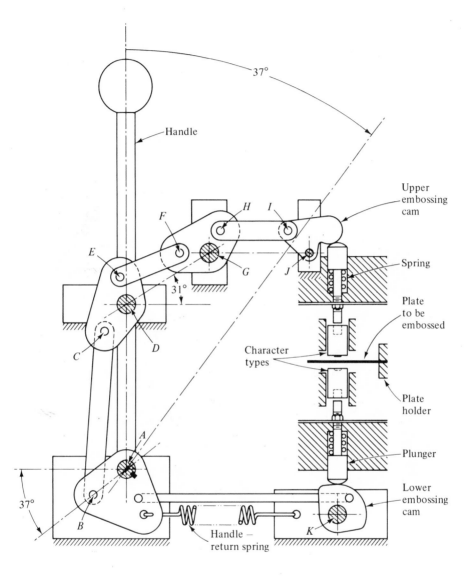

AB = 0.60 in.	FG = 0.38 in.
AD = 2.28 in.	EF = 0.89 in.
DC = 0.47 in.	FH = 0.69 in.
BC = 2.28 in.	GH = 0.41 in.
CE = 0.77 in.	GJ = 1.47 in.
DE = 0.38 in.	JI = 0.38 in.
DG = 1.34 in.	HI = 1.02 in.

Fig. P1.43. *(Cont.)*

41

1.44 Figure P1.44 shows a piston-driven four-bar linkage $ABCD$ used to tilt a cement-mixing drum. Pivot positions are marked in the figure. A, D, and X are fixed ground points which form a straight line. Points B, Y, and C are on one link. Point X is the ground connection for the piston cylinder. Draw a kinematic skeleton of the mechanism when BY is horizontal (loading position). Also, show the linkage configuration in the unloading position. The unloading position is obtained by extending the piston rod 4 ft, $1\frac{1}{2}$ in. from the loading position. Use a scale of 1 ft $= \frac{1}{2}$ in. The dimensions are $YB = 1$ ft, $2\frac{1}{2}$ in.; $BC = 7$ ft, 3 in.; $YC = 7$ ft, 0 in.; $BD = 5$ ft, $1\frac{1}{2}$ in.; $CD = 6$ ft, 5 in.; $DA = 1$ ft, $7\frac{1}{2}$ in.; $BA = 5$ ft, $5\frac{1}{2}$ in.; $ADX = 4$ ft, $10\frac{1}{2}$ in.; and the piston stroke $= 4$ ft, $1\frac{1}{2}$ in.

Fig. P1.44. (Courtesy of Erie Strayer Co.)

Chapter **2**

VECTOR
USAGE

2.1 Scalar Quantities

A quantity which has only magnitude is called a scalar quantity. Such quantities are temperature, time, volume, and density. For example, a room temperature is 70°F. The magnitude 70 is sufficient to specify room temperature; thus temperature is a scalar quantity.

Scalar quantities are mathematically manipulated as ordinary numbers as shown in the following example.

EXAMPLE 2.1

Given: Figure 2.1; a man walks in a straight line from point P to P_3. A clock is started when he begins his walk. Time is recorded at P, P_1, P_2, and P_3.

Determine: (a) Total distance walked. (b) Speed from P to P_3, defined as

$$\text{Speed} = \frac{\text{distance from } P \text{ to } P_3}{\text{time from } P \text{ to } P_3}$$

43

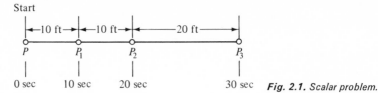

Fig. 2.1. Scalar problem.

Solution. (a) Total distance $= 10 + 10 + 20 = 40$ ft. (b) Speed $= 40$ ft/30 sec $= 1\frac{1}{3}$ ft/sec.

2.2 Introduction to Vector Quantity

Figure 2.2(a) shows two ropes connected to a hook at O. Each rope is being pulled with a 200-lb force. Experimental results show that the two 200-lb forces have the same effect on the hook as one 200-lb force acting vertically upward. The net effect on the hook is shown in Fig. 2.2(b).

The handling of this type of physical problem requires the use of vector mathematics. Simply, vector mathematics deals with vector quantities.

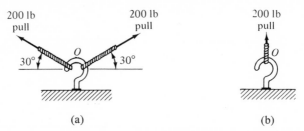

Fig. 2.2. (a) Forces on hook. (b) Net effect on hook.

2.3 Vector Quantities

A vector is defined as a quantity which has magnitude as well as direction. Unfortunately, a quantity with these characteristics may not be a vector. A further restriction imposed on a vector is that it must follow the laws of vector mathematics. More will be said on this in Sect. 2.8.

Examples of vector quantities are force, displacement, velocity, and acceleration.

2.4 Vector Representation

Vector mathematics can be a difficult subject; therefore, we shall limit our discussion primarily to those definitions which can be utilized in this book.

$K_V = 1$ ft/sec/1 unit

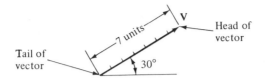

Fig. 2.3. *Graphical repre-sentation of a vector quantity.*

Figure 2.3 shows graphically a vector V which represents a velocity of 7 ft/sec. The scale factor chosen here is $K_V = 1$ ft/sec/1 unit; a distance of 1 unit represents 1 ft/sec. The magnitude of the vector is 7 ft/sec. The direction of the vector is indicated by the arrowhead and the angle 30 deg. The arrowhead end of the vector is called the head and the other extreme the tail of the vector.

Vectors are considered equal if they have the same magnitude and direction; thus in Fig. 2.4, $A = B = C$.

Fig. 2.4. *Equal vectors:* A = B = C.

2.5 Vector Addition

Vector addition of vector A and vector B can be written as

$$A + B = C \qquad\qquad (2.1)$$

It is understood that A, B, and C are vector quantities because the vector plus symbol $+$ is used.

Figure 2.5(a) illustrates the addition of vector B to A. The procedure is to

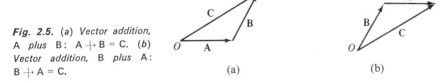

Fig. 2.5. *(a) Vector addition,* A *plus* B: A + B = C. *(b) Vector addition,* B *plus* A: B + A = C.

(a)   (b)

place the tail of vector B to the head of vector A. $A + B$ is equivalent to or has the same effect as C. Vector C is drawn from the origin O, which is the tail of A, to the head of vector B.

Figure 2.5(b) shows that vector C can also be obtained by adding A to B.

This type of relationship is called commutative; that is, $A + B = B + A = C$. The order of adding does not change the outcome.

There are several techniques available for determining a numerical value for vector C. Usually the simplest is the graphical solution. Example 2.2 shows graphically how to solve for vector C.

EXAMPLE 2.2

Given: Velocity vectors A and B; refer to Fig. 2.6(a). $A = 10$ ft/sec, $B = 7$ ft/sec, $\alpha = 25$ deg, and $\beta = 90$ deg.

Determine: Graphically $A + B = C$.

Solution. Velocity will be formally defined in Chapter 4. Let it suffice to say that a point moving at a velocity of 10 ft/sec travels a distance of 10 feet in 1 sec or $\frac{1}{10}$ feet in $\frac{1}{100}$ sec.

A reasonable scale factor for this problem is $K_V = 5$ ft/sec/1 in. If more accuracy is required, a larger scale can be chosen.

STEP 1. Refer to Fig. 2.6(b). Change the velocity to a drawing dimension:

$$\text{Velocity } A = 10 \text{ ft/sec}\left(\frac{1 \text{ in.}}{5 \text{ ft/sec}}\right) = 2 \text{ in.}$$

$$\text{Velocity } B = 7 \text{ ft/sec}\left(\frac{1 \text{ in.}}{5 \text{ ft/sec}}\right) = 1.4 \text{ in.}$$

STEP 2. Start from an arbitrary location O. Place the tail of vector A at O.

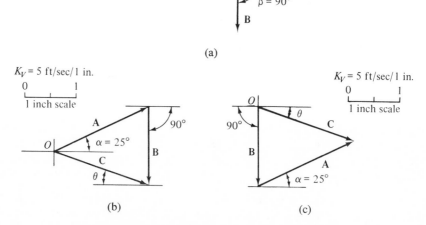

Fig. 2.6. (a) *Given velocities* A *and* B. (b) *Scaled vectors* A + B = C. (c) *Scaled vectors* B + A = C.

STEP 3. Place the tail of B at the head of A.

STEP 4. Vector C is drawn from O to the head of vector B.

STEP 5. Vector C is measured as 1.9 in. long. The velocity C is

$$V_C = 1.9 \text{ in.} \left(\frac{5 \text{ ft/sec}}{1 \text{ in.}} \right) = 9.5 \text{ ft/sec}$$

acting at an angle of $\theta = 17$ deg. Angle θ was measured with a protractor.
Figure 2.6(c) shows that $B +\!\!\!\!+ A$ yields the same result as $A +\!\!\!\!+ B$.

Example 2.2 can be solved mathematically as follows.

EXAMPLE 2.3

Given: Refer to Fig. 2.6(a); velocity $A = 10$ ft/sec, $B = 7$ ft/sec, $\alpha = 25$ deg, and $\beta = 90$ deg.

Determine: Mathematically $A +\!\!\!\!+ B = C$.

Solution. Refer to the vector triangle in Fig. 2.7, labeled OXY:

$$\text{Angle } OXY = 180° - \alpha - \beta = 180° - 25° - 90° = 65°$$

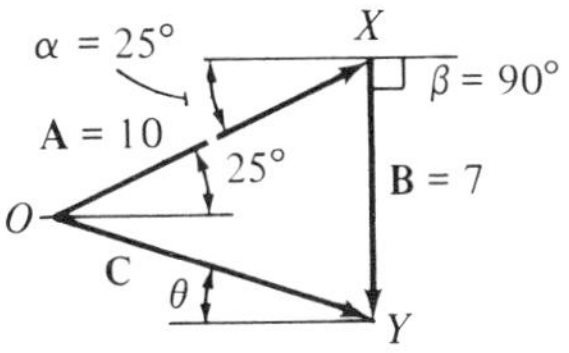

Fig. 2.7. *Vector triangle,* $A +\!\!\!\!+ B = C$.

The cosine law is used to determine vector C:

$$(OY)^2 = (OX)^2 + (XY)^2 - 2(OX)(XY) \cos 65°$$
$$(OY)^2 = (10)^2 + (7)^2 - 2(10)(7)(.423)$$
$$OY = 9.48 \text{ ft/sec}$$
$$\text{Vector } C = 9.48 \text{ ft/sec}$$

The sine law is used to determine angle θ:

$$\frac{OY}{\sin \sphericalangle OXY} = \frac{OX}{\sin \sphericalangle XYO}$$
$$\frac{9.48}{\sin 65°} = \frac{10}{\sin \sphericalangle XYO}$$
$$\sin \sphericalangle XYO = \frac{10}{9.48} \sin 65° = .957, \quad \sphericalangle XYO = 73°$$
$$\theta + 90° - \sphericalangle XYO = 90° - 73° = 17°$$

2.6 Vector Polygon

A common operation is the addition of more than two vectors. The vector configuration formed is called a vector polygon (Figs. 2.8(b) and (c)).

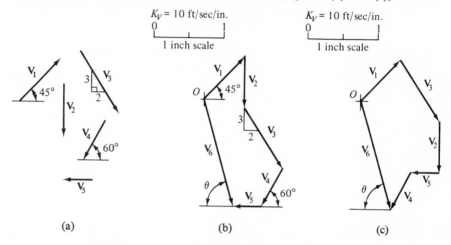

Fig. 2.8. (a) Vectors are in one plane or parallel planes. (b) Vector polygon, $V_1 + V_2 + V_3 + V_4 + V_5 = V_6$. (c) Vector polygon, $V_1 + V_3 + V_2 + V_5 + V_4 = V_6$.

EXAMPLE 2.4

Given: Velocity vectors $V_1, V_2, V_3, V_4,$ and V_5; refer to Fig. 2.8(a). $V_1 = 8$ ft/sec, $V_2 = 7$ ft/sec, $V_3 = 10$ ft/sec, $V_4 = 6$ ft/sec, and $V_5 = 4$ ft/sec.

Determine: Graphically $V_1 + V_2 + V_3 + V_4 + V_5 = V_6$.

Solution. All vectors are scaled proportionately to the scale factor of $K_V = 10$ ft/sec/1 in.

STEP 1. Refer to Fig. 2.8(b). Start addition from O. Place the tail of vector V_1 at O.

STEP 2. Place the tail of V_2 at the head of V_1.

STEP 3. Place the tail of V_3 at the head of V_2.

STEP 4. Continue this "tail-to-head addition" for V_4 and V_5.

STEP 5. Vector V_6 is drawn from O, tail at O, to the head of vector V_5.

STEP 6. V_6 is measured as 1.55 in.:

$$V_6 = 1.55 \text{ in.} \left(\frac{10 \text{ ft/sec}}{1 \text{ in.}}\right) = 15.5 \text{ ft/sec}$$

acting at an angle $\theta = 73.5$ deg.

Figure 2.8(c) illustrates that the sequence in which vectors are added will not change the final result.

The physical significance of this vector addition is that vector V_6 is equivalent or has the same effect as all the vectors $V_1, V_2, V_3, V_4,$ and V_5 acting together.

2.7 Adding Parallel Vectors

The addition of parallel vectors results in a straight-line vector polygon. These parallel vectors can be on one plane or on planes which are parallel to one another. Since all the vectors fall on a straight line, it is easier to handle this as a scalar problem than as a vector problem.

EXAMPLE 2.5

Given: Figure 2.9(a); vertical forces $F_1 = 15$ lb, $F_2 = 20$ lb, $F_3 = 25$ lb, and $F_4 = 40$ lb.

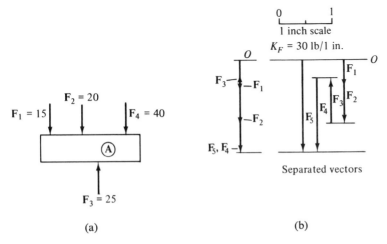

(a) (b)

Fig. 2.9. (a) Forces applied to Ⓐ. (b) Vector polygon, $F_5 = F_1 +\!\!\!+ F_2 +\!\!\!+ F_3 +\!\!\!+ F_4$.

Determine: Total vertical force F_5, where $F_5 = F_1 +\!\!\!+ F_2 +\!\!\!+ F_3 +\!\!\!+ F_4$. Solve (a) as a scalar problem, and (b) as a vector problem.

Solution. (a) Consider downward direction as plus and upward as minus. Then $F_5 = 15 + 20 - 25 + 40 = 50$ lb. A positive value indicates that the 50-lb vector is acting vertically downward. A visual inspection of Fig. 2.9(a) will lead to the same result. The total force down is 75 lb, and the total force up is 25 lb; thus the total force on Ⓐ is 50 lb down. (b) Figure 2.9(b) shows a tail-to-head vector addition of the forces. The vector result is $F_5 = 50$ lb down where F_5 is drawn from O to the head of vector F_4. The vectors are separated in Fig. 2.9(b) so that they can be easily identified.

2.8 Angular Displacement

A quantity having both magnitude and direction need not be a vector quantity. Such a quantity is angular displacement. Angular displacement can be thought of as the angle formed when a link is rotated. Figure 2.10 shows a link

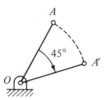

Fig. 2.10. Angular displacement.

OA rotating through an angle of 45 deg. Clearly the magnitude is 45 and the direction is clockwise. Thus angular displacement seems to be a vector quantity. It will be shown in the following discussion that angular displacement does not obey the commutative law of vector addition.

Figure 2.11(a) shows a block, one side of which is darkened. The block is positioned at the origin of the XYZ axis. The XY axes form a plane in the page of the book. The YZ axes form a plane perpendicular to the XY plane. Starting from the position shown in Fig. 2.11(a), the block is rotated 90 deg about the Y axis and then 90 deg about the Z axis, as indicated. The final position is shown in Fig. 2.11(c).

To apply the commutative law, we shall reverse the sequence; that is, first rotate the block about the Z axis and then about the Y axis. Starting from Fig.

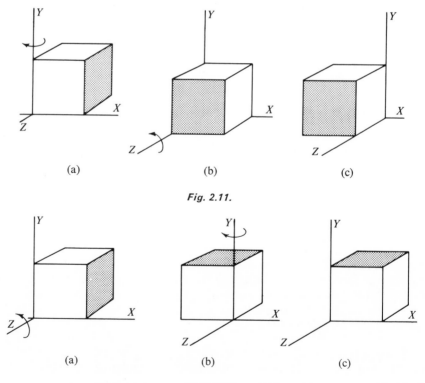

(a)   (b)   (c)

Fig. 2.11.

(a)   (b)   (c)

Fig. 2.12.

2.12(a), which is the same starting position as Fig. 2.11(a), the block is rotated 90 deg about the Z axis and then 90 deg about the Y axis. Comparison of the final positions, Figs. 2.11(c) and 2.12(c), clearly shows that they are different. Obviously the commutative law does not apply to angular displacements; thus angular displacement is not a vector quantity.

2.9 Vector Components

A vector V, shown in Fig. 2.13, can be replaced by many vectors. These vectors are called components of V. The sum of all the components must equal V. If a problem is somewhat confusing, it is sometimes helpful to strike a line through the vector which has been replaced with components.

What is usually desired are two components of the vector V which are 90 deg apart. For example, Fig. 2.14(a) shows a disk $\textcircled{D}$ being pulled at axle O with a force $F = 100$ lb. It is desirable to resolve 100 lb into components, one horizontal and one vertical. These components are shown in Fig. 2.14(b). Horizontal component $F_H = 87$ lb and vertical component $F_V = 50$ lb. Components were graphi-

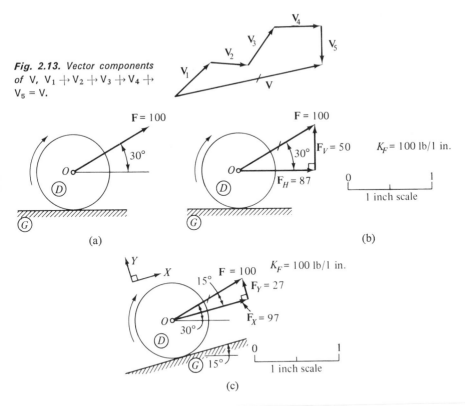

Fig. 2.13. Vector components of V, $V_1 + V_2 + V_3 + V_4 + V_5 = V$.

(a)

(b)

(c)

Fig. 2.14. (a) Disk. (b) $F_H + F_V = F$. (c) $F_X + F_Y = F$.

cally determined. They can also be easily calculated using right-angle trigonometry. From a physical standpoint, F_H and F_V should be applied at axle O.

If the disk were moving on a 15 deg inclined plane, Fig. 2.14(c), the desired components might be those parallel and perpendicular to the plane. These components would be $F_X = 97$ lb and $F_Y = 27$ lb.

EXAMPLE 2.6

Given: Line PQ and velocity P as shown in Fig. 2.15(a).

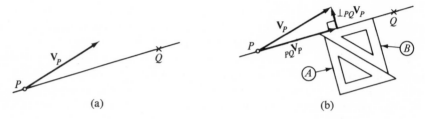

(a)   (b)

Fig. 2.15. (a) *Given* V_P *and line* PQ. (b) $V_P = {}_{PQ}V_P + {}_{\perp PQ}V_P$.

Determine: Velocity components of V_P along line PQ and perpendicular to PQ.

Solution. An explanation of the notation used: V_P is the velocity of P. ${}_{PQ}V_P$ is the velocity of P along line PQ. ${}_{\perp PQ}V_P$ is the velocity of P perpendicular to line PQ. This type of notation will be used in Chapter 4.

STEP 1. Refer to Fig. 2.15(b). Set triangles Ⓐ and Ⓑ so that Ⓑ is lined up with PQ.

STEP 2. Slide triangle Ⓑ upward until a line can be drawn through the tip of arrowhead V_P. This line should be perpendicular to PQ.

STEP 3. Draw component arrowheads in the appropriate direction for $V_P = {}_{PQ}V_P + {}_{\perp PQ}V_P$.

EXAMPLE 2.7

Given: Line CD, ${}_{CD}V_C = 10$ ft/sec and line of action of V_c, as shown in Fig. 2.16(a).

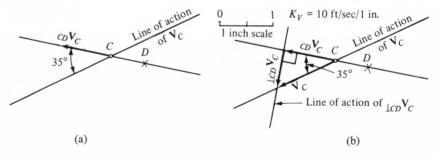

(a)   (b)

Fig. 2.16. (a) *Given* ${}_{CD}V_C$ *and line of action of* V_C. (b) $V_C = {}_{CD}V_C + {}_{\perp CD}V_C$.

Determine: V_C and $_{\perp CD}V_C$.

Solution.

STEP 1. Refer to Fig. 2.16(b). Assume $K_V = 10$ ft/sec/1 in. Layout $_{CD}V_C$ to scale.

STEP 2. Determine $_{\perp CD}V_C$. Draw a line which is perpendicular to CD and also passes through the tip of arrowhead $_{CD}V_C$.

STEP 3. Complete the vector triangle. $V_C = {}_{CD}V_C \mathbin{+\!\!\!\rightarrow} {}_{\perp CD}V_C$.

STEP 4. Scale off V_C and $_{\perp CD}V_C$. $V_C = 12.3$ ft/sec and $_{\perp CD}V_C = 7$ ft/sec.

2.10 Vector Subtraction

Subtraction of vector B from A is defined as

$$A \longrightarrow B = C \tag{2.2}$$

or

$$A \mathbin{+\!\!\!\rightarrow} (-B) = C \tag{2.3}$$

In Eq. (2.2), vector subtraction is symbolized as $\longrightarrow$, whereas in Eq. (2.3), subtraction is accomplished by the addition of a negative vector. A negative vector is obtained by directing a positive vector in the opposite direction. Equation (2.3) will be employed for vector subtraction rather than Eq. (2.2) because the rules of vector addition have already been explained. The direct use of Eq. (2.2) would require a new set of rules.

To illustrate vector subtraction, assume vectors A and B are given [Fig. 2.17(a)] and determine $A \longrightarrow B$. Vectors A and B are considered positive. Using

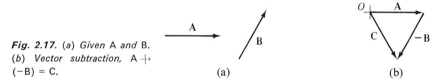

Fig. 2.17. (a) Given A and B. (b) Vector subtraction, $A \mathbin{+\!\!\!\rightarrow} (-B) = C$.

(a)

(b)

the form $A \mathbin{+\!\!\!\rightarrow} (-B)$, the procedure is to reverse the arrowhead of B to obtain $-B$ and then to place the tail of $-B$ at the head of A. Figure 2.17(b) shows the result of this addition. Vector C is directed from O, tail at O, to the head of $-B$.

It is interesting to note that in a few instances a vector equation can be handled similarly to an algebraic equation. Compare the following vector and algebraic forms (the vector form is on the left and the algebraic form is on the right):

Vector form	*Algebraic form*
Case 1	
Given (1) $A \longrightarrow B = C$	*Given* (1) $A - B = C$
can be changed to	can be changed to
(2) $A \mathbin{+\!\!\!\rightarrow} (-B) = C$	(2) $A + (-B) = C$

Vector Eq. (2) shows that $+\!\!\!\rightarrow(-B) = \longrightarrow B$; this resembles algebraic Eq. (2) where $+(-B) = -B$.

Case 2

Given (3) $A \longrightarrow B = C$	*Given* (3) $A - B = C$
can be changed to	can be changed to
(4) $A \longrightarrow C = B$	(4) $A - C = B$

Here the rule of "transposition" was used.

Case 3

Given (5) $A +\!\!\!\rightarrow B = C$	*Given* (5) $A + B = C$
can be changed to	can be changed to
(6) $B = C \longrightarrow A$	(6) $B = C - A$

Again the rule of "transposition" was used.

It must be emphasized that the three preceding examples were presented only as a memory aid. Generally, vector equations are not manipulated as algebraic equations.

The examples which follow serve to show the use of vector subtraction. Vector subtraction is used to determine the change of a vector quantity and the relative vector. Both of these topics will be discussed in detail in later chapters.

EXAMPLE 2.8

Given: Oscillating mechanism shown in Fig. 2.18(a). At position 1, the slider is moving at a velocity of 8 ft/sec; call this V_1. When the slider reaches position 2, the velocity is $V_2 = 12$ ft/sec.

Determine: Change in velocity from position 1 to position 2. Change in velocity is defined as $\Delta V = V_2 \longrightarrow V_1$, where ΔV is treated as one quantity.

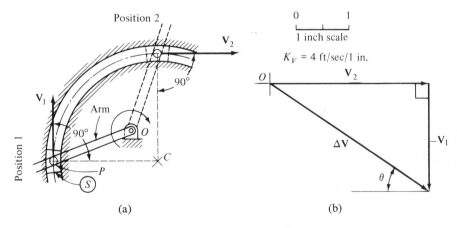

Fig. 2.18. (a) Given V_1 and V_2. (b) Vector subtraction, $\Delta V = V_2 +\!\!\!\rightarrow (-V_1)$.

Solution. Slider Ⓢ moves in a circular path about point C. A pin P is attached to the slider. As the arm rotates, the pin slides along the arm. Distance OP is constantly changing.

$\Delta V = V_2 \longrightarrow V_1$ is read as "delta V equals V two minus V one." The Δ symbol is often used to indicate a change or difference in a quantity.

The completed solution is shown in Fig. 2.18(b). Use the vector equation $\Delta V = V_2 +\!\!\!\!\!\!\!\!\!+ (-V_1)$. Reverse the arrowhead of V_1 to obtain $-V_1$. Place the tail of $-V_1$ at the head of V_2. ΔV is the vector from O to the head of $-V_1$. Scale off ΔV. $\Delta V = 14.4$ ft/sec, acting at an angle of $\theta = 33.6$ deg.

EXAMPLE 2.9

Given: Two cars A and B approaching an intersection. Car A is moving at 30 mph and B at 40 mph. Refer to Fig. 2.19(a).

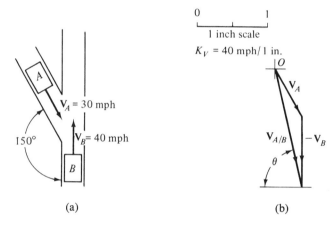

(a)   (b)

Fig. 2.19. (a) Cars A and B approaching intersection. (b) Vector subtraction, $V_{A/B} = V_A +\!\!\!\!\!\!\!\!\!+ (-V_B)$.

Determine: Relative velocity of car A with respect to car B. Relative velocity is defined as $V_{A/B} = V_A \longrightarrow V_B$.

Solution. $V_{A/B}$ is read as "velocity of A relative to B." This problem can be interpreted to mean "How fast is car A approaching car B?"

The completed solution is shown in Fig. 2.19(b). Use the vector equation $V_{A/B} = V_A +\!\!\!\!\!\!\!\!\!+ (-V_B)$. Add $-V_B$ to V_A. $V_{A/B} = 68$ mph at an angle of $\theta = 77.5$ deg.

2.11 Multiplication and Division of a Vector by a Scalar

A vector which is multiplied or divided by a scalar quantity is still a vector quantity. For example, Fig. 2.20 shows an acceleration vector $A = 40$ ft/sec^2 which is multiplied by 2 sec and divided by 2 sec. The resulting quantities V and J are vectors, where

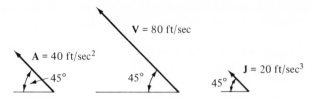

Fig. 2.20. *Multiplication and division of acceleration vector A by a scalar quantity.*

$$V = 2A = 2 \sec \left(40 \frac{ft}{\sec^2} \right) = 80 \text{ ft/sec}$$

and

$$J = \frac{A}{2} = \frac{40 \text{ ft/sec}^2}{2 \sec} = 20 \text{ ft/sec}^3$$

PROBLEMS

2.1 A man walks in a straight line through points A, B, and C. $AB = 10$ ft and $BC = 10$ ft. It takes him 5 sec to move from A to B and 5 sec to move from B to C. He pauses 5 sec at point B. Determine (a) the total distance traveled, (b) the speed from point A to B (do not consider 5 sec pause), (c) the speed from point B to C (do not consider 5 sec pause), and (d) the speed from point A to C.

2.2 A car travels the first lap around a 1-mile track at 1 mile/min. The second lap is at the rate of $\frac{1}{2}$ mile/min. What is the combined speed for both laps?

2.3 Check Fig. 2.2(a) experimentally using 200 g instead of 200 lb, connect three spring scales together at appropriate angles. All scales should indicate 200 g. Mount on a wooden board.

2.4 Graphically determine the vector sum for A and B (Fig. P2.4). The magnitudes of A and B both equal 10 ft/sec.

Fig. P2.4.

2.5 Do Prob. 2.4 mathematically.

2.6 In Fig. P2.6, add acceleration vector A_1 to A_2. $A_1 = 10$ ft/sec² and $A_2 = 15$ ft/sec².

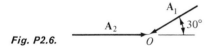

Fig. P2.6.

2.7 Figure P2.7 shows the line of actions of velocity V_1, V_2, and V_3. Determine graphically V_3 if $V_3 = V_2 \mathbin{+\!\!\!+} V_1$, where $V_1 = 100$ ft/sec at an angle of 30 deg, as shown. A line of action is defined as a line, or a line parallel to the given line, on which the vector acts.

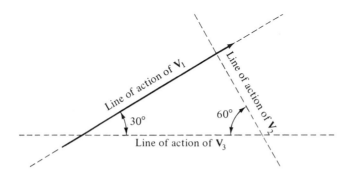

Fig. P2.7.

2.8 In Fig. P2.8, the line of action of V_A is known. Determine V_C if $V_A \mathbin{+\!\!\!+} V_C = V_E$, where $V_E = 10$ ft/min.

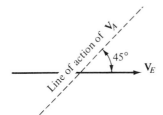

Fig. P2.8.

2.9 Solve Prob. 2.8 with the additional information that $V_A = 20$ ft/min. Is there more than one answer?

2.10 Figure P2.10 shows a C-shaped link rotating clockwise about O. Acceleration of point P is defined as $A_P = A_1 + A_2$, where $A_1 = 10$ ft/sec² and $A_2 = 4$ ft/sec². Determine A_P.

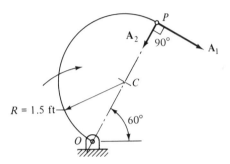

Fig. P2.10.

2.11 Figure P2.11 shows part of the velocity analysis of a slider crank. $V_B = 10$ ft/sec and $V_X = 7.2$ ft/sec. Graphically determine V_C for the position shown, where $V_C = V_X + V_B$. $AB = 1$ in. and $BC = 5$ in.

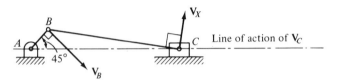

Fig. P2.11.

2.12 Figure P2.12 shows vector forces A, B, C, and D, where $A = 50$ lb, $B = 40$ lb, $C = 45$ lb, and $D = 35$ lb. Graphically determine vector sum $A + B + C + D$.

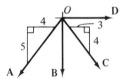

Fig. P2.12.

2.13 In Fig. P2.13, determine the vector sum for displacement vectors D_1 to D_6. Vectors D_2 and D_5 are horizontal, and D_4 and D_6 are vertical. $D_1 = 1$ ft, $D_2 = 3$ ft, $D_3 = 1$ ft, $D_4 = 2$ ft, $D_5 = 3$ ft, and $D_6 = 2$ ft.

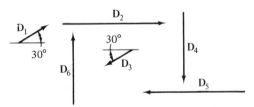

Fig. P2.13.

2.14 Figure P2.14 shows scaled vectors A, B, and C. Show graphically that $A \mathbin{+\!\!\!\!+} B \mathbin{+\!\!\!\!+} C = D$ can be solved as $A \mathbin{+\!\!\!\!+} B = X$ and $X \mathbin{+\!\!\!\!+} C = D$.

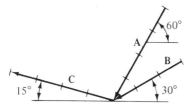

Fig. P2.14.

2.15 In Fig. P2.15, a man walks on a flat surface from A to D, passing points B and C along the way. Determine (a) the total distance walked, and (b) the displacement vector AD, where $AD = AB \mathbin{+\!\!\!\!+} BC \mathbin{+\!\!\!\!+} CD$.

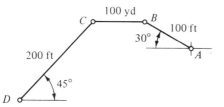

Fig. P2.15.

2.16 In Fig. P2.16, $V_A = 1.4$ in./sec, $V_B = 1.1$ in./sec, $V_D = 1.0$ in./sec, and $V_E = 2.75$ in./sec. Determine V_C. The adding order recommended is $V_A \mathbin{+\!\!\!\!+} V_B \mathbin{+\!\!\!\!+} V_D \mathbin{+\!\!\!\!+} V_C = V_E$.

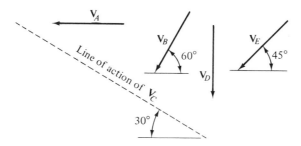

Fig. P2.16.

2.17 Figure P2.17 shows the line of actions for accelerations A_B, A_C, and A_D. A_A is directed vertically downward. Determine A_D, where $A_D = A_A \mathbin{+\!\!\!\!+} A_B \mathbin{+\!\!\!\!+} A_C$.

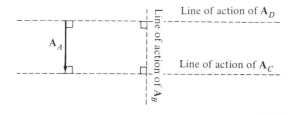

Fig. P2.17.

2.18 Figure P2.18 shows a block acted on by four forces. Determine the result of $F_1 + F_2 + F_3 + F_4$. Will the block "feel" these forces? Discuss.

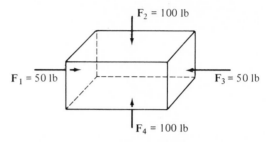

Fig. P2.18.

2.19 In Fig. P2.19, a point P is located in a three-dimensional object, where $S = S_Z + S_X + S_Y$, $S_Z = 1.0$ in., $S_X = .5$ in., and $S_Y = .75$ in. Determine S completely. *Hint:* $S_Z + S_X = S_W$ in plane XZ and $S_W + S_Y = S$ in plane YOP.

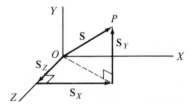

Fig. P2.19.

2.20 A group of scaled parallel vectors is shown in Fig. P2.20. Determine the result of $A + B + C + D$.

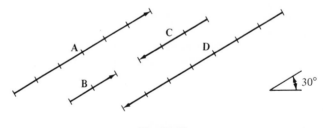

Fig. P2.20.

2.21 Collinear vectors are added as $V_1 + V_2 + V_3 = V_4$. $V_1 = V_2 = 10$ ft/hr and $V_4 = 20$ ft/hr. V_4 is opposite in sense to V_1. Determine V_3 in inches per second.

2.22 Figure P2.22 shows three parallel planes. $V_A = 5$ ft/sec, $V_B = 3$ ft/sec, and $V_C = 1$ ft/sec. Determine the vector sum $V_A + V_B + V_C$.

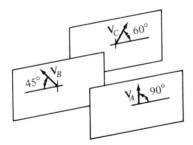

Fig. P2.22.

2.23 Figure P2.23 shows a block, one side of which is darkened. The block is rotated 90 deg clockwise about the Z axis and then 180 deg counterclockwise about the Z axis. Determine the final position of the block. Is the commutative law valid here?

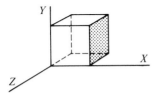

Fig. P2.23.

2.24 In Fig. P2.24, $V_Q = 40$ ft/sec. Determine the components of V_Q along line PQ and perpendicular to PQ.

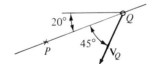

Fig. P2.24.

2.25 In Fig. P2.25, a force $F = 10$ lb acts on lever L at point P. Graphically determine components of F parallel and perpendicular to PQ. Link $PQ = 1$ ft.

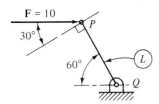

Fig. P2.25.

2.26 Solve Prob. 2.25 mathematically.

2.27 Figure P2.27 shows a parallelogram configuration. Show that A and B are components of C.

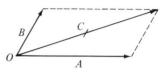

Fig. P2.27.

2.28 In Fig. P2.28, a triangular link ABC rotates about point C. The velocity of point A is $V_A = 10$ ft/sec. Determine the components of V_A parallel and perpendicular to AB.

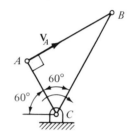

Fig. P2.28.

2.29 In Fig. P2.29, a block is being pulled with a force $F = 50$ lb. Show that F_1 is *not* the maximum component of F in the horizontal direction.

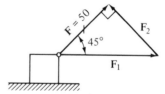

Fig. P2.29.

2.30 Figure P2.30 shows $_{\perp AB}V_A$, the line of action of V_A, and line AB. If $_{\perp AB}V_A = 15$ ft/sec, determine V_A. Use the vector equation $V_A = {}_{\perp AB}V_A + {}_{AB}V_A$.

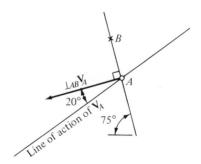

Fig. P2.30.

2.31 Figure P2.31 shows a four-bar linkage $ABCD$, where $V_B = 3$ ft/sec, $AB = 1$ in., $BC = 3\frac{1}{2}$ in., $CD = 2$ in., and $DA = 3$ in. For the position shown, determine V_C. Use the following equations in the sequence shown:

(1) $V_B = {}_{BC}V_B + {}_{\perp BC}V_B$ (construct at point B)

(2) ${}_{BC}V_B = {}_{BC}V_C$ (construct at point C)

(3) $V_C = {}_{BC}V_C + {}_{\perp BC}V_C$ (construct at point C)

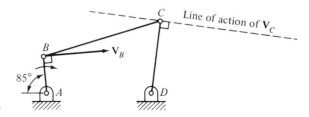

Fig. P2.31.

2.32 In Fig. P2.32, $V_A = 10$ in./sec and $V_B = 20$ in./sec. Determine $V_B \rightarrow V_A$.

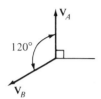

Fig. P2.32.

2.33 In Fig. P2.33, point B has a velocity of $V_1 = 12.5$ ft/sec at position 1 and $V_2 = 15$ ft/sec at position 2. Determine the change in velocity, defined as $\Delta V = V_2 \rightarrow V_1$.

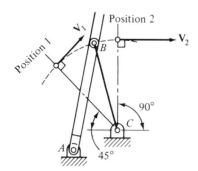

Fig. P2.33.

2.34 Show that the commutative law is applicable to $A + (-B) = C$.

2.35 For the scaled vectors shown in Fig. P2.35, determine vector E. E is defined as $E = A + B + C \longrightarrow D$.

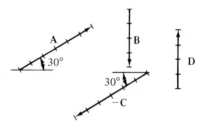

Fig. P2.35.

2.36 Each gear of a crossed helical pair has a surface velocity of 500 ft/min, as shown in Fig. P2.36. Determine the sliding velocity between gears. The sliding velocity is defined as $V = V_2 \longrightarrow V_1$.

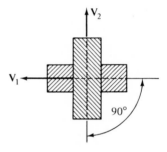

Fig. P2.36.

2.37 In Fig. P2.37, cars A, B, and C approach an intersection. $V_A = 30$ mph, $V_B = 40$ mph, and $V_C = 30$ mph. Determine $V_{B/A} = V_B \longrightarrow V_A$ and $V_{B/C} = V_B \longrightarrow V_C$.

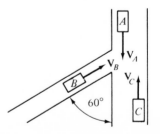

Fig. P2.37.

2.38 At the instant shown in Fig. P2.38, $V_A = 20$ ft/sec and $V_B = 25$ ft/sec. Link dimension $\overline{AB} = 3$ in. Determine the velocity of B relative to A, which is defined as $V_{B/A} = V_B \longrightarrow V_A$. Also, determine $V_{A/B}$. Is there any similarity between $V_{B/A}$ and $V_{A/B}$?

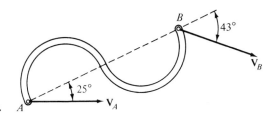

Fig. P2.38.

2.39 Figure P2.39 shows a point P on disk $\textcircled{D}$. The acceleration of P relative to the stationary ground is $A_P = 10$ ft/sec^2, and acceleration $A_O = 0$ ft/sec^2. Determine $A_{O/P}$, where $A_{O/P} = A_O \longrightarrow A_P$.

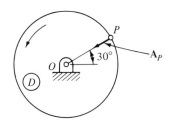

Fig. P2.39.

2.40 Divide the vector velocity .004 ft/sec by a scalar quantity of .00002 sec. Is the answer a vector quantity?

Chapter **3**

MOTION CONCEPT

3.1 Relative Motion

A link (or body) is in motion when after a time interval a measurable difference in position, from the starting position, is observed with respect to or relative to a reference. To help visualize motion relative to a reference link, imagine yourself fixed onto the reference link observing the moving link.

For example, Fig. 3.1 concerns the positions of links at times 0, 5, and 10 sec. Figure 3.1(a) shows links $\textcircled{A}$ and $\textcircled{B}$ connected to frame $\textcircled{F}$ at points X and Y. Frame $\textcircled{F}$ is connected to $\textcircled{E}$ at point Z.

Comparison of Fig. 3.1(b) with Fig. 3.1(a) shows that $\textcircled{F}$, as well as $\textcircled{A}$ and $\textcircled{B}$, moved 10 deg clockwise relative to $\textcircled{E}$. Note that $\textcircled{A}$ has not moved relative to $\textcircled{B}$. If $\textcircled{A}$ has not moved relative to $\textcircled{B}$, then $\textcircled{B}$ has not moved relative to $\textcircled{A}$.

Comparison of Fig. 3.1(c) with Fig. 3.1(b) shows that $\textcircled{A}$ and $\textcircled{B}$ moved 90 deg counterclockwise relative to $\textcircled{F}$ or relative to $\textcircled{E}$. $\textcircled{F}$ has not moved relative to $\textcircled{E}$. It is interesting to note that $\textcircled{A}$ has moved relative to $\textcircled{B}$. In Fig. 3.1(b), $\textcircled{A}$ is positioned directly above $\textcircled{B}$; imagine yourself fixed onto $\textcircled{B}$. Figure

66

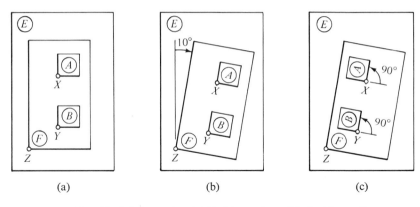

(a)   (b)   (c)

Fig. 3.1. *Relative motion.* (a) *Position at 0 sec.* (b) *Position at 5 sec.* (c) *Position at 10 sec.*

3.1(c) shows A positioned to the right of B. Thus A has moved relative to B.

The earth's surface is frequently chosen as a stationary or fixed reference. Motion relative to the earth is conventionally called absolute motion. Motion relative to a moving body is conventionally called relative motion. For mechanism purposes it is usually convenient to think of a mechanism frame as fixed to the earth. Thus link motion relative to the frame is absolute.

The terms *absolute* and *relative* will be used frequently in conjunction with angular and linear displacement, angular and linear velocity, and angular and linear acceleration.

3.2 Plane Motion

This text is primarily concerned with plane motion. A link which moves in a plane or parallel to a plane is said to be in plane motion. Links shown in Fig. 3.1 were in plane motion. An eraser which is moved on a blackboard is in plane motion.

Plane motion is classified into three groups:

1. Translation
2. Rotation
3. Combination

3.3 Plane Translation

A link is in translation when a straight line on the link (actual or imaginary) moves parallel to its initial position.

Figures 3.2(a) and 3.2(b) show link L translating; line XY is always parallel

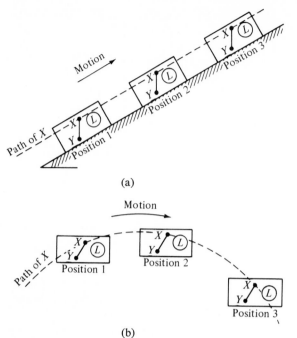

(a)

(b)

Fig. 3.2. (a) Plane rectilinear translation. (b) Plane curvilinear translation.

to position 1. For rectilinear translation, point X moves in a straight path. For curvilinear translation, X moves in a curved path.

3.4 Plane Rotation

A link is in rotation when a straight line on the link (actual or imaginary) moves in a circular path about a point. The line stays a fixed distance from the point of rotation.

In Fig. 3.3 link L is rotating about point C. Note that XY is a constant distance from C and that points X and Y move in circular arcs about point C.

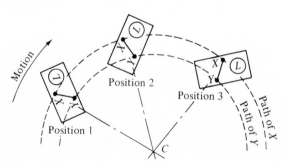

Fig. 3.3. Plane rotation about C.

It is important to realize that a point cannot rotate (only a body or line can rotate) because theoretically a point has no size; it is infinitesimally small. Thus it is impossible to determine an angle of rotation. In Fig. 3.4 a point Y is represented as moving from Y' to Y'''. All that can be said about the motion of Y is that it is translating—specifically, curvilinear translation.

Fig. 3.4. Path of point Y.

We often say that a point is rotating. What is really meant is that a small body, which we conveniently call a point, is rotating. For example, Fig. 3.5 shows a grinding wheel rotating at 1800 rpm. A point P (actually particle P is meant) is on the periphery of the wheel. Technically it is not correct to say that point P rotates at 1800 rpm. A better statement would be that line PO rotates at 1800 rpm or particle P rotates at 1800 rpm.

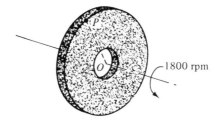

Fig. 3.5. Grinding wheel rotates at 1800 rpm.

3.5 Plane Combined Motion

A link which moves with neither plane translation nor rotation is moving with combined motion. Combined motion is a combination of translation and rotation.

The slider crank, Fig. 3.6(a), is a good example of a mechanism which exhibits all three types of plane motion. Slider Ⓢ moves with rectilinear translation, crank AB rotates about A, and link BC moves with combined motion from position BC to $B'C'$. The motion of link BC can be separated into rotation and translation or translation and rotation. To illustrate these component motions it is necessary to disconnect the pin connections at B and C. Starting from the initial position BC, Fig. 3.6(b), rotate link BC about C to B_1C; now translate the link to the final position $B'C'$. Figure 3.6(c) shows the reverse sequence for BC; that is, translate then rotate.

(a)

(b)

(c)

Fig. 3.6. (a) Slider crank. (b) Combined motion; rotate then translate. (c) Combined motion; translate then rotate.

3.6 Radian

A radian is an angular measurement, just as a degree is an angular measurement, defined as

$$\frac{\text{Arc length}}{\text{Radius}} = \text{radians}$$

For example, in Fig. 3.7 arc length $AB = .8$ in. and the radius $OA = OB = 1.53$ in., the radian value of θ is

$$\frac{\text{Arc } AB}{\text{Radius}} = \frac{.8 \text{ in.}}{1.53 \text{ in.}} = .523 \text{ rad}$$

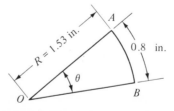

Fig. 3.7. $\theta = .523$ rad.

The radian is a dimensionless quantity, which means that both the numerator and denominator are expressed in the same units.

It is useful to know the relationship between radians and degrees. Figure 3.8 shows a circle of radius R. There are 360 deg around the center O. To determine the number of radians around O, use the radian definition:

$$\frac{\text{Arc length}}{\text{Radius}} = \frac{\text{circumference}}{\text{radius}} = \frac{2\pi R}{R} = 2\pi \text{ rad}$$

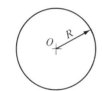

Fig. 3.8. *Center angle, 360° = 2π rad.*

Therefore

$$360° = 2\pi \text{ rad}$$

or, dividing by 2, we obtain

$$180° = \pi \text{ rad}$$

EXAMPLE 3.1

Given: A lathe chuck rotates at 60 rpm.

Determine: The rotational speed in radians per second.

Solution. One revolution is defined as 360 deg; therefore 1 rev = 360 deg = 2π rad:

$$60\frac{\text{rev}}{\text{min}}\left(\frac{1 \text{ min}}{60 \text{ sec}}\right)\left(\frac{2\pi \text{ rad}}{1 \text{ rev}}\right) = \frac{60(2\pi)}{60}\frac{\text{rad}}{\text{sec}} = 2\pi\frac{\text{rad}}{\text{sec}}$$

3.7 Angular Displacement

The angular displacement of a link is the angle which a straight line on the link (actual or imaginary) sweeps through. It is a scalar quantity which can be either clockwise or counterclockwise. Units for angular displacement are degrees, radians, etc.

In Fig. 3.9, link $\textcircled{L}$ is one link of a mechanism (mechanism not shown) which moves from position XY to $X'Y'$. XY is a line on $\textcircled{L}$. The path of $\textcircled{L}$ cannot be determined from Fig. 3.9; thus it is impossible to label $\textcircled{L}$ as rotating

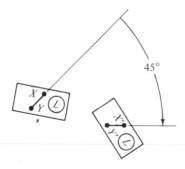

Fig. 3.9. *Angular displacement of $\textcircled{L}$ or* XY *is 45 deg cw.*

or moving with combined motion. Regardless, the angular displacement of L or XY is 45 deg cw.

Figures 3.10(a) and 3.10(b) illustrate possible mechanism arrangements for

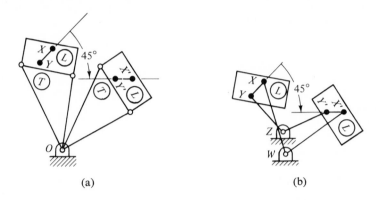

(a)

(b)

Fig. 3.10. (a) L or XY rotates 45 deg cw about O. (b) L or XY moves in combined motion 45 deg cw.

the link L shown in Fig. 3.9. In Fig. 3.10(a), L is rotating about O, and link T is rigidly attached to L. In Fig. 3.10(b), L is the coupler link of a four-bar linkage $WXYZ$. Here L is in combined motion. Note that in both figures the angular displacement is 45 deg cw.

The coupler link BC, Fig. 3.11(a), of a four-bar linkage $ABCD$ moves from

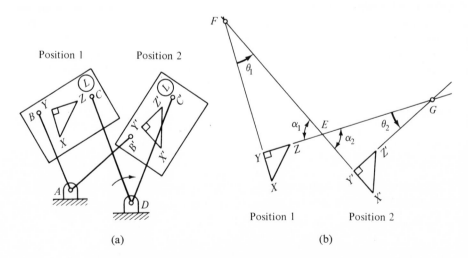

(a)

(b)

Fig. 3.11. (a) Four-bar linkage; triangle XYZ marked onto coupler coupler link L. (b) Positions 1 and 2 of triangle XYZ.

position 1 to position 2. It can be shown that all the lines on link BC move through the same angle. To help prove this statement we shall mark triangle XYZ onto coupler link BC. Triangle XYZ moves from XYZ to $X'Y'Z'$. To simplify matters, the two positions for triangle XYZ have been redrawn in Fig. 3.11(b). Proof will be shown only for lines XY and YZ. The angular displacement of XY is θ_1 ccw, and for YZ it is θ_2 ccw. The following sequence shows that the magnitude of $\theta_1 = \theta_2$:

1. $\alpha_1 = \alpha_2$; opposite angles are equal.
2. $\sphericalangle FYE = \sphericalangle EY'G$; right angles are equal.
3. $\theta_1 = \theta_2$; third angle of similar triangles FYE and $EY'G$ must be equal.

For angular displacements to be considered equal, both magnitude and direction of rotation must be the same. This is true here; thus the angular displacement of XY is equal to YZ.

EXAMPLE 3.2

Given: Figure 3.12(a); the center of gear $\textcircled{D}$, which is point C, translates a distance of π in. The pitch radius of the gear is 1 in.

Determine: Angular displacement for line XY. Line XY is marked on the gear.

Solution. This problem can be treated as a disk (radius, 1 in.) which rolls, without slip, on a flat stationary surface [Fig. 3.12(b)]. To help visualize the motion, a length of π in. is marked off on the disk's circumference and on the stationary surface. Both of these lengths are divided into six equal parts and are labeled 0 to 6. Note that π in. is half of the disk's circumference. As the disk $\textcircled{D}$ rolls from left to right, point 1 on $\textcircled{D}$ will eventually correspond with point 1 on $\textcircled{S}$. Later, point 2 on $\textcircled{D}$ corresponds with point 2 on $\textcircled{S}$, etc. Observe that line $C6$, at the start of motion, is vertical with point 6 above C. In the final position, $C6$ is again vertical, but 6 is below C. Line $C6$ has an angular displacement of 180 deg cw. Since all lines on a rigid body move through the same angle, for plane motion, it can be concluded that XY has an angular displacement of 180 deg cw.

The method used to solve Example 3.2 is obviously lengthy; however, it does bring insight into the solution. Actually, to solve this particular type of problem, all that is required is the conversion of an arc length into an angular value. The arc length is the distance the disk rolls. Example 3.3 illustrates this type of solution.

EXAMPLE 3.3

Given: A gear and stationary rack. The gear center translates a distance of 10 in. The pitch radius of the gear is 1 in.

Determine: Angular displacement for the gear.

Solution. The gear's circumference is $2\pi(1) = 2\pi$ in.

$$10 \text{ in.} \left(\frac{1 \text{ circumference}}{2\pi \text{ in.}}\right)\left(\frac{360°}{1 \text{ circumference}}\right) = 574°$$

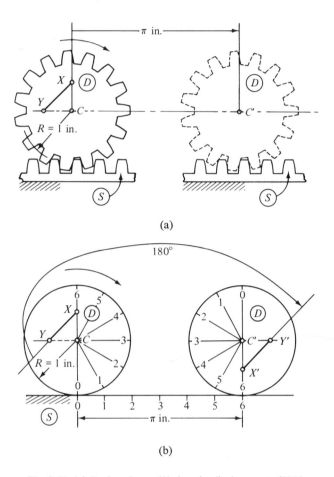

(a)

(b)

Fig. 3.12. (a) Rack and gear. (b) Angular displacement of XY is 180 deg cw.

For drafting use it may be more convenient to use an angle of 214 deg, 574 deg − 360 deg = 214 deg, rather than 574 deg. To be precise, the term *angular displacement* refers to an angle less than 360 deg; thus the answer to this example would be, angular displacement is 214 deg. The total angle or angular distance is 574 deg.

In Examples 3.2 and 3.3 the motion of the gears are classified as combined motion. It is understood that this motion is relative to the stationary rack or earth. Relative to another body, the motion may not be classified as combined motion. Figure 3.13(a), which is practically identical to Fig. 3.12(a), shows a slider Ⓛ connected to gear Ⓓ at point *C*. The slider is held between stationary guides Ⓖ. Relative to the stationary rack Ⓢ or guides Ⓖ, gear Ⓓ moves with combined motion, and Ⓛ moves with rectilinear translation. Figure 3.13(b)

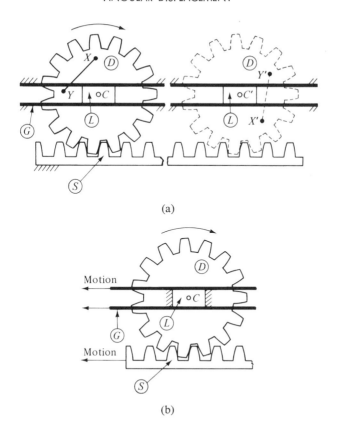

(a)

(b)

Fig. 3.13. (a) L translates, and D moves with combined motion relative to S or G. (b) S and G translate, and D rotates relative to L.

shows the motion as it would appear relative to L. Gear D rotates cw about C, and S and G translate to the left.

EXAMPLE 3.4

Given: Figure 3.14; the center of planet gear P moves an arc distance of $\frac{3}{2}\pi$ in. Pitch radius for planet P is 1 in., and the pitch radius for the stationary sun gear S is 2 in.

Determine: The angular displacement for line XY on gear P.

Solution. This combination of gears is called a planetary gear system. Arm A holds the gears in mesh. To simplify the analysis, gears P and S will be treated as disks. Point C, on the arm, translates a circular arc distance of $\frac{3}{2}\pi$ in. Thus the arm moves through an angle of

$$\frac{\text{Arc}}{\text{Radius}} = \frac{\frac{3}{2}\pi \text{ in.}}{3 \text{ in.}} = \frac{\pi}{2} \text{ rad} \quad \text{or } 90°$$

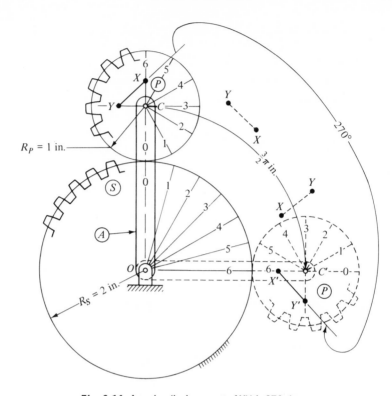

Fig. 3.14. Angular displacement of XY *is 270 deg cw.*

Contacting surfaces on ⓢ and ⓟ are divided into six equal parts and are labeled 0 to 6. During motion the numbers on ⓟ correspond to the same numbers on ⓢ. To help visualize the motion of XY, two intermediate positions are shown. Carefully observe that the angular displacement for line $C6$ is 270 deg cw; therefore XY moves 270 deg cw.

3.8 Relative Angular Movement

Angular displacement is always measured with respect to another link or body. Absolute angular displacement makes reference to a stationary link, whereas relative angular displacement is used with respect to a moving link. The term *angular displacement* by itself is usually assumed to mean absolute angular displacement.

In Fig. 3.15(a), line XY is marked on link Ⓛ. Ⓛ or XY moves through an angular displacement of 45 deg cw with respect to the background link Ⓜ. This angular displacement is symbolized as $\theta_{L/M}$ (read as theta L relative to M). For ease of writing, the link symbol ◯ is not employed in the subscript of θ.

Figure 3.15(b) shows that link Ⓜ, shown in Fig. 3.15(a), is not stationary but rotates 75 deg cw, about point O, relative to link Ⓔ. While this motion is

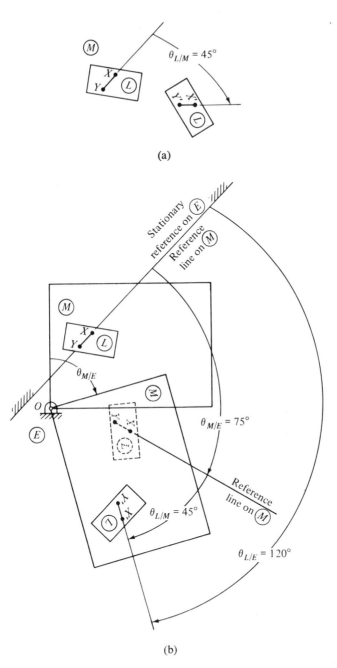

(a)

(b)

Fig. 3.15. (a) Angular displacement of (L) relative to (M), $\theta_{L/M} = 45$ deg cw. (b) Angular displacement of (L) relative to (E), $\theta_{L/E} = 120$ deg cw.

77

taking place, link L moves from position XY to $X'Y'$. There are two motions to contend with: L relative to M, and M relative to E. Reference lines are established on E and M from which $\theta_{L/M}$, $\theta_{M/E}$, and $\theta_{L/E}$ can be measured. A convenient location for the stationary reference line on E, as well as the line on M, is collinear to position XY (which is the starting position of L). It can be observed that the absolute angular displacement of L relative to E, $\theta_{L/E}$, is 120 deg cw.

The following relative equations can be written for plane angular displacement [refer to Fig. 3.15(b)]:

$$\theta_{L/E} = \theta_{L/M} + \theta_{M/E} \qquad (3.1)$$

$$\theta_{L/M} = \theta_{L/E} - \theta_{M/E} \qquad (3.2)$$

$$\theta_{M/E} = \theta_{L/E} - \theta_{L/M} \qquad (3.3)$$

Here these equations are algebraic equations, and the rules of algebra apply. To use these equations properly it is necessary to establish a sign convention, for example, plus for cw and minus for ccw.

Let us apply these equations to a problem. Assume that $\theta_{L/M} = 45$ deg cw and $\theta_{M/E} = 75$ deg cw, as in Fig. 3.15(b). Find $\theta_{L/E}$. Using Eq. (3.1) and assuming cw is plus,

$$\theta_{L/E} = \theta_{L/M} + \theta_{M/E} = 45° + 75° = 120° \text{ cw}$$

The relative vector equations for linear displacement, which will be established in Sect. 3.10, are very similar in form to the relative algebraic equations for angular displacement just developed. It is important that we be able to write these equations quickly from memory.

For example, a problem involves links A, B, and C, where A is a stationary link (earth). Determine $\theta_{B/A}$. Referring to Eq. (3.1),

$$\theta_{B/A} = \theta_{B/C} + \theta_{C/A}$$

In another example, links C, D, and E are involved, where E is the stationary link. Establish $\theta_{C/D}$. Referring to Eq. (3.2),

$$\theta_{C/D} = \theta_{C/E} - \theta_{D/E} \qquad (1)$$

It is possible to solve this problem starting with Eq. (3.1). Using Eq. (3.1) as reference, we have

$$\theta_{C/D} = \theta_{C/E} + \theta_{E/D} \qquad (2)$$

This is a perfectly valid equation; however, it will be altered so that it looks exactly like Eq. (1). Since $\theta_{E/D} = -\theta_{D/E}$, Eq. (2) can be written as

$$\theta_{C/D} = \theta_{C/E} - \theta_{D/E}$$

This is exactly the same as Eq. (1). The statement was made that $\theta_{E/D} = -\theta_{D/E}$. Physically this means that if link (E) moves 10 deg cw relative to (D), then (D) moves 10 deg ccw relative to (E).

EXAMPLE 3.5

Given: Figure 3.16(a); arm (A) rotates 90 deg cw. Pitch radius for planet (P) is 1 in., and the pitch radius for the stationary sun gear (S) is 2 in.

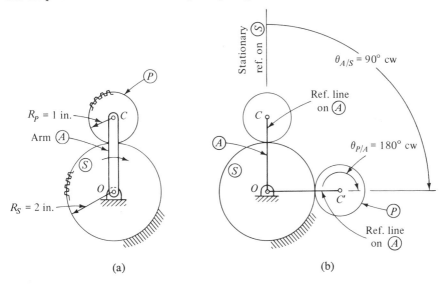

(a)   (b)

Fig. 3.16. (a) Planetary. (b) Angular displacement of (P) relative to (S), $\theta_{P/S} = 270$ deg cw.

Determine: Angular displacement of (P) relative to (S) using the relative angular displacement equation.

Solution. This example is the same as Example 3.4. Involved in this example are bodies (A), (P), and (S). Determine $\theta_{P/S}$. Referring to Eq. (3.1),

$$\theta_{P/S} = \theta_{P/A} + \theta_{A/S} \tag{1}$$

To evaluate $\theta_{P/A}$ and $\theta_{A/S}$, two reference lines must be established, one on (A) and the other on (S). The arm, line OC in Fig. 3.16(b), will serve as the reference line on (A) (note, line OC moves). The stationary reference line, on body (S), is collinear to position OC.

Gear (P) meshes with the stationary sun (S) for a circumferential distance of

$$\tfrac{1}{4}[2\pi R_S] = \tfrac{1}{4}[2\pi(2)] = \pi \text{ in.}$$

Gear $\textcircled{P}$ relative to $\textcircled{A}$ is

$$\theta_{P/A} = \pi \text{ in.} \left[\frac{360°}{2\pi(1)} \right] = 180° \text{ cw}$$

The angular displacement of $\textcircled{A}$ relative to $\textcircled{S}$ was given as 90 deg cw; therefore, $\theta_{A/S} = 90$ deg cw. Evaluate $\theta_{P/S}$. Substitute into Eq. (1) and assume that cw is plus:

$$\theta_{P/S} = 180° + 90° = 270° \text{ cw}$$

3.9 Linear Displacement

In Fig. 3.17, a point P moves along path 1 through positions P' and P''. Point O is a fixed point, which was arbitrarily chosen. The position or location of P' and P'' from O are represented by position vectors S_1 and S_2.

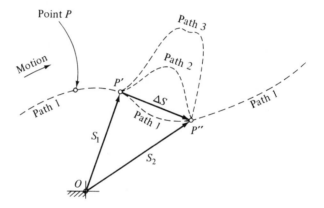

Fig. 3.17. Change in position vector, $\Delta S = S_2 \to S_1$.

The change in position of point P when moving from P' to P'' is defined as a vector which extends from P' to P''. This vector is labeled ΔS. Symbol ΔS is read as delta S, where ΔS is considered as one quantity. ΔS is directed from P' toward P'' and the magnitude is the straight-line distance $P'P''$ measured in feet, inches, etc. This is not the actual distance traveled by point P. Had P traveled along path 2 or 3 when moving from P' to P'', the change in position vector ΔS would still be the same. Changing the position of fixed point O will not alter ΔS.

The vector triangle shown in Fig. 3.17 can be written as

$$S_1 + \Delta S = S_2 \tag{3.4}$$

Rearranging Eq. (3.4),

$$\Delta S = S_2 \to S_1 \tag{3.5}$$

If we assume the reference point O to be located at position P', Eq. (3.5) reduces to $\Delta S = S_2$ because $S_1 = 0$. In this case, S_2 or simply S is commonly referred to as the linear displacement.

EXAMPLE 3.6

Given: Roller Ⓡ, Fig. 3.18, moves between stationary guides.

Determine: Linear displacement for point C from C' to C''.

Solution. Locate the stationary reference point at position C', Fig. 3.18. The linear displacement from C' to C'' is vector S.

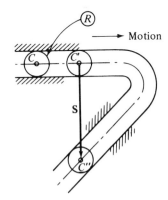

Fig. 3.18. *Vector* S *is the linear displacement from* C' *to* C''.

3.10 Relative Linear Movement

Linear displacement, like angular displacement, is either absolute or relative. The term *linear displacement* by itself is assumed to mean absolute linear displacement.

Figure 3.19(a) shows a link Ⓛ, pivoted at P, driving roller Ⓡ around a grooved plate Ⓖ. Point C is the center of Ⓡ. The linear displacement of point C from C' to C'' is shown as vector $S_{C/R}$. Reference point R is located on Ⓖ at position C'.

Figure 3.19(b) shows that as C moves from C' to C'', plate Ⓖ rotates cw about point O relative to Ⓔ. The absolute displacement of point C, when moving from C' to C'', is labeled $S_{C/E}$. Linear displacement of reference point R relative to Ⓔ is $S_{R/E}$. The relative vector equations are, referring to Fig. 3.19(b),

$$S_{C/E} = S_{R/E} + S_{C/R}$$

or

$$S_{C/E} = S_{C/R} + S_{R/E} \tag{3.6}$$

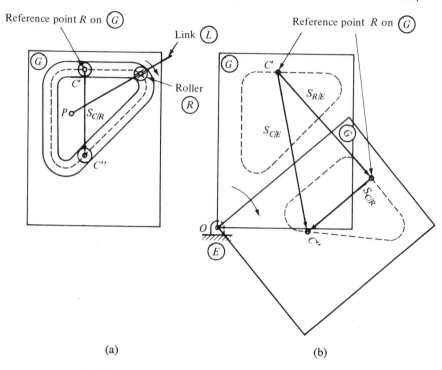

(a)   (b)

Fig. 3.19. (a) Displacement of point C from C' to C'' relative to ⓖ, $S_{C/R}$. Reference R is established on ⓖ. (b) Displacement of point C from C' to C'' relative to ⓔ, $S_{C/E}$. E is established on earth.

$$S_{C/R} = S_{C/E} \to S_{R/E} \tag{3.7}$$

$$S_{R/E} = S_{C/E} \to S_{C/R} \tag{3.8}$$

Reference point R, if required, is established at the initial position of displacement.

EXAMPLE 3.7

Given: Figure 3.20(a); a man walks 10 ft on a boat deck from position M' to M''. During the same time period the boat moves 100 ft forward.

Determine: Absolute linear displacement of the man.

Solution. In this example the man and the boat are considered as points. Symbols to be used are:

$S_{M/E}$ = displacement of the man relative to the earth
$S_{M/B}$ = displacement of the man relative to the boat
$S_{B/E}$ = displacement of the boat relative to the earth

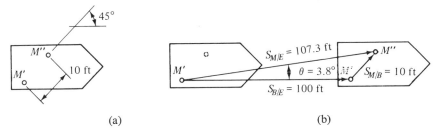

(a)   (b)

Fig. 3.20. (a) The man's displacement relative to the boat.
(b) Displacement of the man relative to Ⓔ, $S_{M/E} = S_{M/B} +\!\!> S_{B/E}$.

Use the vector equation

$$S_{M/E} = S_{M/B} +\!\!> S_{B/E}$$
$$S_{M/E} = 10 +\!\!> 100 = 107.3 \text{ ft at } \theta = 3.8 \text{ deg}$$

The graphical solution is shown in Fig. 3.20(b).

PROBLEMS

3.1 As shown in Fig. P3.1, blocks Ⓐ and Ⓒ translate to the right and Ⓑ to the left relative to the earth Ⓔ. (a) What is the absolute motion of Ⓐ, Ⓑ, Ⓒ, and Ⓔ? (b) What is the motion of Ⓐ relative to Ⓑ? (c) What is the motion of Ⓑ relative to Ⓐ?

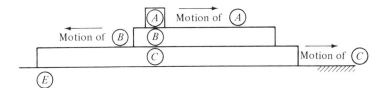

Fig. P3.1.

3.2 In Fig. P3.2, pin Ⓟ is fastened securely to wheel Ⓦ. What is the absolute motion of Ⓦ, Ⓛ, and Ⓟ?

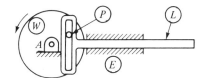

Fig. P3.2.

3.3 For a parallel four-bar linkage $ABCD$, what is the motion of BC relative to AD and BC relative to AB?

3.4 In the slider crank ABC, Fig. P3.4, AD is one link. What is the motion of links AB, BC, and Ⓢ relative to AD?

Fig. P3.4.

3.5 Refer to Fig. P3.4. What is the motion of links AD, AB, and Ⓢ relative to BC?

3.6 Clock Ⓐ is positioned upside down in Fig. P3.6(a). In Fig. P3.6(b), clock Ⓑ is upside down and is also turned around. Determine the motion of the hour hand Ⓧ relative to the hour hand Ⓨ in both figures.

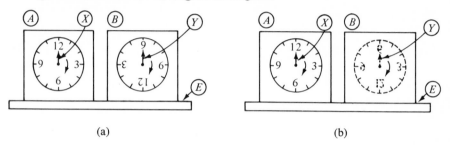

(a)   (b)

Fig. P3.6.

3.7 Knowing that 180 deg $= \pi$ rad, determine the number of degrees in 1 rad.

3.8 A circular segment has an arc of 6 in. and a radius of 2 ft. Determine the included angle in degrees.

3.9 A circular segment has a cord length of .7 in. and a radius of 1.0 in. Determine the included angle in radians and the included arc length in inches.

3.10 A link rotates at $30/\pi$ rpm. Determine angular velocity in radians per second.

3.11 In Fig. P3.11, Ⓦ rotates cw at 1 rad/sec. Determine the angle, in degrees, that XY moves through in two seconds.

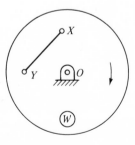

Fig. P3.11.

3.12 Link AB of a four-bar linkage moves 45 deg cw from the position shown in Fig. P3.12. Determine the angular displacement of BC, CD, and CE. $AB = .70$ in., $BC = 3.25$ in., $CD = 1.00$ in., $CE = 1.50$ in., and $EA = 3.00$ in.

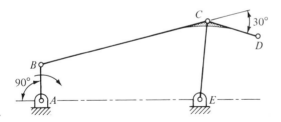

Fig. P3.12.

3.13 Starting from the position shown in Fig. P3.12, determine the angular displacement of BC, CD, and CE if AB moves 93.5 deg cw.

3.14 Refer to Fig. 3.12(a). Center C on gear $\textcircled{D}$ translates 2π in. Determine the angular displacement for XY. Also, show that $\textcircled{D}$ moves with combined motion by separating the motion into translation and rotation. Assume that $R = 1$ in.

3.15 Figure P3.15 shows a simplified sketch of part of a roller printer machine. The hand-operated carriage houses a pulley which is keyed to a rubber roller. When the carriage is moved, the pulley is forced to turn, thus turning the roller. Determine the angular displacement of the roller for a carriage displacement of 1 in. The pulley diameter is 2 in.

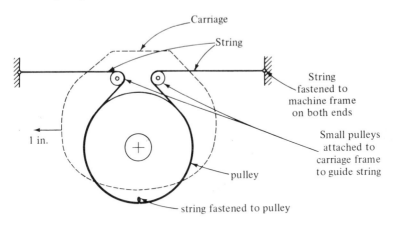

Fig. P3.15. (Courtesy of Pitney-Bowes, Inc.)

3.16 In Fig. P3.16, gear $Ⓐ$ moves through an angular displacement of 90 deg cw. Determine the angular displacement for $Ⓑ$ and $Ⓛ$. $XY = 3$ in.

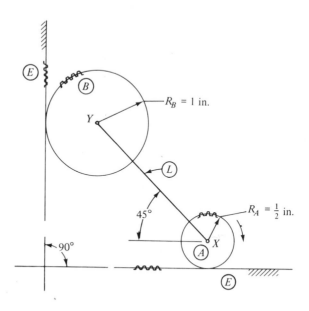

Fig. P3.16.

3.17 Figure P3.17 shows a toy with link BC attached to the wheel $Ⓦ$. Pivot pins are fixed onto the frame $Ⓕ$ at points A and D. DCE is one link. Wheel $Ⓦ$ rolls on $Ⓔ$ without slipping. At the instant shown in the figure, AB is horizontal. The dimensions are $AB = \frac{1}{4}$ in., $BC = 3\frac{3}{4}$ in., $DC = \frac{9}{16}$ in., $DCE = 4$ in., $AD = 3\frac{5}{8}$ in., and the wheel radius is $\frac{3}{4}$ in. What type of motion has DCE relative to the earth? Relative to the frame? If the frame moves 2 in. to the right, what is the angular displacement for DCE? Where should the drum be positioned?

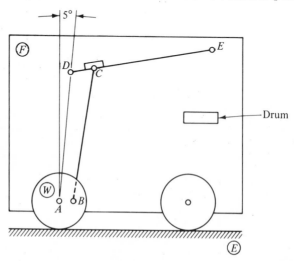

Fig. P3.17.

3.18 Use the same gear radii shown in Fig. 3.14. Rotate arm Ⓐ 150 deg cw about point O. The sun gear is stationary. Determine the absolute angular displacement for the planet gear. To help visualize this motion, label a radial line on Ⓟ and plot the position of this line for every 30 deg of arm rotation.

3.19 Use the same gear radii shown in Fig. 3.14. Rotate arm Ⓐ 150 deg ccw about point C. Gear Ⓟ is stationary. Determine the angular displacement of gear Ⓢ relative to Ⓟ. To help visualize this motion, label a radial line on Ⓢ and plot the position of this line for every 30 deg of arm rotation.

3.20 Use the same gear radii shown in Fig. 3.14. Keep arm Ⓐ stationary, and let Ⓟ and Ⓢ be free to move. If Ⓟ rotates 150 deg cw, what is the angular displacement of Ⓢ?

3.21 In Fig. P3.21, arm Ⓐ rotates 45 deg cw about point B. Gear Ⓕ is stationary. The pitch radii are Ⓕ, 1 in.; Ⓖ, $\frac{1}{2}$ in.; and Ⓗ, $\frac{3}{4}$ in. Determine the absolute angular displacement for Ⓖ and Ⓗ. The graphical method is required.

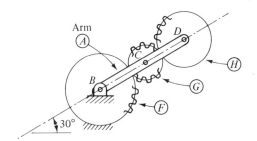

Fig. P3.21.

3.22 In Fig. P3.22, gear Ⓖ rotates 30 deg cw about point B. At the instant shown, link DE is horizontal. Determine the angular displacement for Ⓗ. The graphical method is required. DE is 1 in.

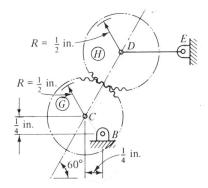

Fig. P3.22.

3.23 In Fig. P3.23, gear G is attached to slider S at O. G is free to rotate about O. Rack M is free to move in either direction. $AB = AC = .50$ in., $OB = 3.10$ in., and $DC = 2.85$ in. If link BAC rotates 60 deg cw, determine the angular displacement for G.

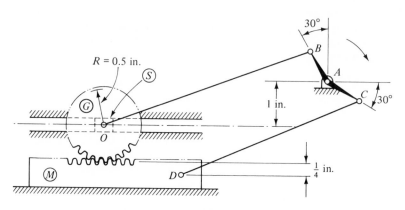

Fig. P3.23.

3.24 Figure P3.24 shows a Bourdon gage used to measure pressure. The pointer is rigidly fastened to gear P. BC is part of sector S. Pivot pins are fixed onto the gage at points A and B. At the instant shown, link CD lies along the 68-deg line, and A and D are on the same horizontal line. The dimensions are $DC = \frac{37}{64}$ in., $BC = \frac{25}{64}$ in., the pitch radius for gear P is $\frac{1}{4}$ in., and the pitch radius for S is $\frac{3}{4}$ in. The pressure in the tube causes point D to move to D', along the 68-deg line; $DD' = \frac{1}{4}$ in. Determine the angular displacement of the pointer.

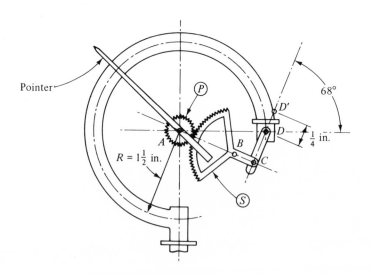

Fig. P3.24.

3.25 An angular displacement problem involves three bodies Ⓐ, Ⓑ, and Ⓒ. Write the angular equations for Ⓒ relative to Ⓐ, Ⓐ relative to Ⓒ, and Ⓒ relative to Ⓑ.

3.26 Two identical coins, with good serrated edges, are required for this problem (Fig. P3.26). Glue the flat surface of one coin onto a piece of cardboard. Without slipping, roll the other coin cw around the stationary coin for a contacted distance of one circumference. Determine experimentally and by calculation the angular rotation for the moving coin.

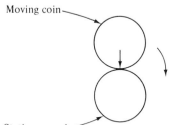

Moving coin

Fig. P3.26. Stationary coin

3.27 Determine the answer to Prob. 3.19 using a relative angular equation.

3.28 Determine the answers to Prob. 3.21 using the relative angular equations. Also, calculate the angular displacement of Ⓗ relative to Ⓖ.

3.29 After a torque is applied to gear Ⓒ, Fig. P3.29, it is found that the angular displacement of gear Ⓐ relative to Ⓑ is .020 rad cw, Ⓑ relative to Ⓒ is .025 rad cw, and Ⓓ relative to Ⓒ is .055 rad cw. Determine $\theta_{A/C}$, $\theta_{A/D}$, and $\theta_{D/A}$. View the shaft from the right end.

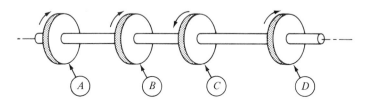

A B C D

Fig. P3.29.

3.30 The compound shaft shown in Fig. P3.30 is subjected to applied torques at faces B and C. Face A is securely attached to a solid support. The angular displacement of face B relative to face A is .005 rad ccw, and face C relative to B is .015 rad cw. Determine the angular displacement of face C relative to face A.

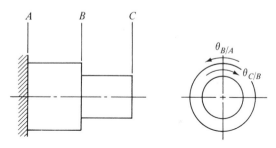

Fig. P3.30.

3.31 In Fig. P3.31, a cylinder is mounted parallel to gear racks $\widehat{M}$ and $\widehat{S}$. Rack $\widehat{S}$ is stationary. For a piston stroke of 2 in., determine the linear displacement for $\widehat{M}$. The gear radius is 1 in. The graphical method is required.

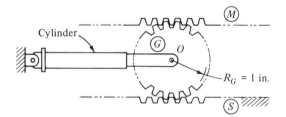

Fig. P3.31.

3.32 For the mechanism shown in Fig. P3.32, determine the linear displacement for point P. The cam rotates 60 deg cw from the position shown. The length $XY = 5$ in.

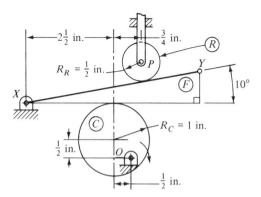

Fig. P3.32.

3.33 Using the same information given in Example 3.7, determine the linear displacement of the earth relative to the man.

Chapter **4**

VELOCITY

4.1 Introduction

A velocity analysis of a mechanism is important for determining timing, wear, acceleration, torque, force, and energy. For example, the surface wear of a cam and its follower is related to sliding velocity. The kinetic energy or energy of motion of a punch press ram is related to the ram velocity.

A velocity analysis helps the designer visualize the motion of a mechanism over a period of time or at an instant in time.

4.2 Average Angular Velocity

Angular velocity is defined as the change in angular displacement divided by the change in time; thus

$$\text{Angular velocity} = \frac{\text{change in angular displacement}}{\text{change in time}} \tag{4.1}$$

For plane motion, the angular velocity is cw or ccw depending on the direction of angular displacement. Units for angular velocity are rad/sec, rad/min, rev/min, etc.

There are two categories of angular velocity:

1. Average angular velocity
2. Instantaneous angular velocity

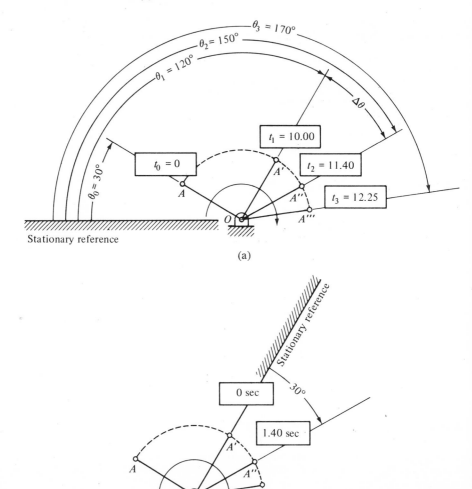

(a)

(b)

Fig. 4.1. (a) *Angular displacement is measured in degrees from a stationary reference. Time is recorded from position* OA *in seconds.* (b) *The same motion in Fig. 4.1.* (a). *Angular displacement is measured from a stationary reference located at* OA'. *Time is recorded from* OA'.

The average is for a relatively long period of time, whereas the instantaneous is for an extremely short or instant period of time.

Using Eq. (4.1), the average angular velocity equation is written as

$$\omega_{av} = \frac{\Delta\theta}{\Delta t} \tag{4.2}$$

The symbol ω (omega) will be used to identify angular velocity. Figure 4.1(a) shows a link OA rotating, without stopping, from starting position OA to OA'''. We can imagine, as link OA rotates, that pictures are taken with a high-speed camera at positions OA', OA'', and OA'''. Figure 4.1(a) shows the superimposing of these pictures. The angular positions of link OA are measured from a horizontal stationary reference. A timing device is started at position OA, and time is recorded at positions OA', OA'', and OA'''. The average angular velocity from position OA' to OA'' is

$$\omega_{av} = \frac{\Delta\theta}{\Delta t} = \frac{\theta_2 - \theta_1}{t_2 - t_1} = \frac{(150 - 120)°}{(11.40 - 10.00)\ \text{sec}} = \frac{30°}{1.40\ \text{sec}} = 21.43°/\text{sec cw}$$

or, changing to radians per second,

$$\omega_{av} = 21.43°/\text{sec} \left(\frac{\pi\ \text{rad}}{180°}\right) = .37\ \text{rad/sec cw}$$

Note that ω_{av} is cw; this is the same direction as $\Delta\theta$.

The average angular velocity of OA between OA' and OA'' could have been determined from Fig. 4.1(b). The motion shown in Fig. 4.1(b) is exactly the same as that shown in Fig. 4.1(a); however, the stationary reference is now located at OA' and time is recorded from OA'. Utilizing the data shown in Fig. 4.1(b),

$$\omega_{av} = \frac{\Delta\theta}{\Delta t} = \frac{30°}{1.40\ \text{sec}} = 21.43°/\text{sec cw}$$

4.3 Instantaneous Angular Velocity

Figure 4.2, similar to Fig. 4.1(b), shows additional positions of link OA between OA' and OA''. The average angular velocities are determined for the following positions:

$$OA' \text{ to } OA'': \quad \omega_{av} = \frac{30°}{1.40\ \text{sec}} = 21.43°/\text{sec}$$

$$OA' \text{ to } OA_1: \quad \omega_{av} = \frac{20°}{.95\ \text{sec}} = 21.05°/\text{sec}$$

$$OA' \text{ to } OA_2: \quad \omega_{av} = \frac{10°}{.49\ \text{sec}} = 20.41°/\text{sec}$$

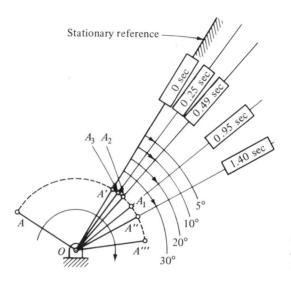

Stationary reference

0 sec
0.25 sec
0.49 sec
0.95 sec
1.40 sec

A_3 A_2

A'
A_1
A''
A'''

5°
10°
20°
30°

Fig. 4.2. Similar to Fig. 4.1. (b). Additional positions have been added between OA' and OA''.

$$OA' \text{ to } OA_3: \quad \omega_{av} = \frac{5°}{.25 \text{ sec}} = 20.00°/\text{sec}$$

Imagine that there are 1000 positions of link OA between OA' and OA'' with OA_{1000} being closest to OA'. The average angular velocity between OA' and OA_{1000} would be almost the same as the instantaneous angular velocity at position OA'.

The instantaneous angular velocity is obtained by having the change in time Δt, as well as the change in angular displacement $\Delta\theta$, approach zero. Instantaneous angular velocity is often expressed as

$$\omega = \frac{d\theta}{dt} \tag{4.3}$$

The expression $d\theta/dt$ is used in the mathematics of calculus. Another useful calculus expression is $\omega = \lim_{\Delta t \to 0} \Delta\theta/\Delta t$, which means that the instantaneous angular velocity equals the limit of the average quantity $\Delta\theta/\Delta t$ as time approaches zero.

In Eq. (4.3), $d\theta$ (d theta) and dt (d tee) are considered infinitesimal quantities. Although $d\theta$ and dt are infinitesimal, the instantaneous angular velocity is not necessarily small. For example,

$$\frac{d\theta}{dt} = \frac{.00000000000000000020°}{.00000000000000000001 \text{ sec}} = \frac{20 \times 10^{-21}}{1 \times 10^{-21}} °/\text{sec} = 20°/\text{sec}$$

In summary, the instantaneous angular velocity, as its name implies, is the angular velocity at an instant in time. The term *angular velocity* by itself will usually mean instantaneous angular velocity, or, to be exact, absolute instantaneous angular velocity.

4.4 Angular Velocity

In Sect. 3.7 it was shown that lines on a link or body which move with plane motion move through the same angular displacement. The angular displacement ($\Delta\theta$) for each line takes place during the same span of time (Δt). Thus the average angular velocity ($\omega_{av} = \Delta\theta/\Delta t$) for each line as well as the average angular velocity for the body are equal. Assume that $\Delta\theta$ and Δt are infinitesimal quantities $d\theta$ and dt. For plane motion only, the instantaneous angular velocity ($\omega = d\theta/dt$) for the body and for the lines on the body are equal.

The reader should realize that the instantaneous angular velocity—or, simply, the angular velocity—is a vector quantity. It can be shown that the infinitesimal $d\theta$ is a vector quantity. (This is seemingly a paradox because θ is a scalar quantity [Sect. 2.8].) Accepting the fact that $d\theta$ is a vector quantity, ω must also be a vector quantity because $\omega = d\theta/dt$, and a vector divided by a scalar is a vector. The angular velocity vector acts perpendicular to the plane on which motion takes place. For plane motion mechanisms, the angular velocity vectors are parallel or collinear to one another. For simplicity these vectors will be treated as scalar quantities. This simplification cannot be applied to all mechanisms; space mechanisms, such as the gyroscope, require that angular velocity be treated as a vector quantity.

The angular velocity of a link or a line on a link can be approximately determined by applying the definition $\omega \approx \omega_{av} = \Delta\theta/\Delta t$, where $\Delta\theta$ and Δt are small quantities. Although this technique is time consuming and is generally not used for determining ω, it is valuable because of its simplicity.

EXAMPLE 4.1

Given: Figure 4.3(a); crank AB of slider crank ABC rotates at a constant 2 rad/sec cw. $AB = BC = BD = DC = 2$ in.

Determine: Approximate angular velocity of line BD when $\theta = 30$ deg. Use $\omega \approx \omega_{av} = \Delta\theta/\Delta t$, where $\Delta\theta$ and Δt are assumed to be small quantities.

Solution. To solve this problem we must assume a mechanism size and an angular displacement for driver link AB.

STEP 1. For the completed solution, refer to Fig. 4.3(b). Assume that $K_M = 1$ in./1 in., and let $\Delta\theta_{AB} = .1$ rad. The time required to move .1 rad is

$$\Delta t = \frac{\Delta\theta_{AB}}{\omega_{AB}} = \frac{.1 \text{ rad}}{2 \text{ rad/sec}} = .05 \text{ sec}$$

STEP 2. To construct $\Delta\theta_{AB} = .1$ rad we shall use a radius of 5 in. and an arc length of $\frac{1}{2}$ in. Extend line AB. Measure off a 5 in. distance from A along line AB; this locates X. Using point A as center, draw arc 1 through X.

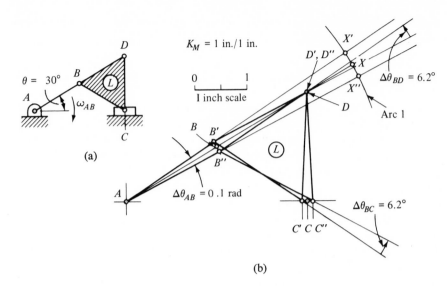

(a)

(b)

Fig. 4.3. (a) *Slider crank* ABC. (b) *Angular velocity of line* BD *when at position* BD, $\omega_{BD} \approx \Delta\theta_{BD}/\Delta t$.

STEP 3. Measure off a $\frac{1}{4}$ in. chord distance on both sides of X; this locates X' and X''. Draw in lines AX' and AX''. Construction shows that $\Delta\theta_{AB} = .1$ rad:

$$\Delta\theta_{AB} = \frac{\text{arc } X'XX''}{AX} = \frac{\frac{1}{2}\text{ in.}}{5\text{ in.}} = .1 \text{ rad}$$

This is not exactly correct because a $\frac{1}{2}$ in. chord was substituted for a $\frac{1}{2}$ in. arc.

STEP 4. Locate points B', C', and D' and then B'', C'', and D''.

STEP 5. Measure the angular displacement between $B'D'$ and $B''D''$. $\Delta\theta_{BD} = 6.2$ deg ccw; also note that $\Delta\theta_{BC} = 6.2$ deg ccw.

STEP 6. Determine ω_{BD}:

$$\omega_{BD} \approx \frac{\Delta\theta_{BD}}{\Delta t} = \frac{6.2°}{.05 \text{ sec}} = 124°/\text{sec ccw} \quad \text{or} \quad 2.16 \text{ rad/sec ccw}$$

(The exact answer, determined from another method, is $\omega_{BD} = 2$ rad/sec ccw).

The angular velocity for lines on link L are equal; namely, $\omega_{BD} = \omega_{BC} = \omega_{DC} = \omega_L \approx 2.16$ rad/sec ccw. It is interesting to note that in Fig. 4.3(b), D' and D'' are practically on D. Point D is called the instant center of rotation for link L when L is at position BCD. It appears as if link L is rotating about D at this instant. More will be said on instant center later in this chapter.

In Example 4.1 we determined the approximate absolute angular velocity. Example 4.2 concerns relative angular velocity.

EXAMPLE 4.2

Given: Figure 4.4 shows two positions of a linkage. The time between positions is 2 sec.

Determine: Approximate angular velocity of link ⓕ relative to ④. Use the information shown in Fig. 4.4.

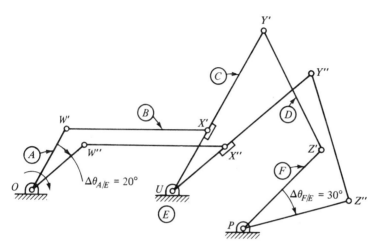

Fig. 4.4.

Solution. Links ⓕ, ④, and ⓔ are involved; hence

$$\Delta\theta_{F/A} = \Delta\theta_{F/E} - \Delta\theta_{A/E}$$

Refer to Fig. 4.4, assuming cw is plus:

$$\Delta\theta_{F/A} = 30° - 20° = 10° \text{ cw} \quad \text{or} \quad .175 \text{ rad cw}$$

$$\omega_{F/A} \approx \frac{\Delta\theta_{F/A}}{\Delta t} = \frac{.175 \text{ rad}}{2 \text{ sec}} = .087 \frac{\text{rad}}{\text{sec}} \text{ cw}$$

A general approach to this example would be to use the relative angular velocity equation:

$$\omega_{F/A} = \omega_{F/E} \rightarrow \omega_{A/E}$$

For plane motion this equation can be written in algebraic form as

$$\omega_{F/A} = \omega_{F/E} - \omega_{A/E}$$

Since we can only approximate, use the form

$$\omega_{F/A} \approx \frac{\Delta\theta_{F/E}}{\Delta t_1} - \frac{\Delta\theta_{A/E}}{\Delta t_2}$$

where $\Delta t_1 = \Delta t_2 = 2$ sec and cw is plus:

$$\omega_{F/A} \approx \frac{30°}{2} - \frac{20°}{2} = 5°/\text{sec} = .087 \frac{\text{rad}}{\text{sec}} \text{ cw}$$

Note that since the angles are large it is doubtful if the answer is representative of the true relative angular velocity.

4.5 Linear Velocity

It is frequently necessary to determine the linear velocity of a point. At times a link (or body), which consists of many points or particles, can be analyzed as a point. For example, an automobile, a projectile, and a ball in motion are frequently treated as a point in motion.

Average linear velocity is the change in linear displacement divided by the change in time:

$$V_{av} = \frac{\Delta S}{\Delta t} \tag{4.4}$$

If Δt and ΔS are considered to be infinitesimals dt and dS, then the instantaneous linear velocity is

$$V = \frac{dS}{dt} \tag{4.5}$$

Both average and instantaneous linear velocity are vector quantities, where the velocity vector points in the same direction as the linear displacement. Typical units for linear velocity are ft/sec, in./min, and mi/hr. Linear velocity is commonly referred to as velocity.

The term *velocity* should not be confused with the term *speed*. Speed is a scalar quantity; it is the total distance moved divided by time.

EXAMPLE 4.3

Given: Same information as Example 4.1; refer to Fig. 4.3(a).

Determine: Approximate instantaneous linear velocity of point C when $\theta = 30$ deg. Use $V \approx V_{av} = \Delta S/\Delta t$; ΔS and Δt are small quantities.

Solution. This example is presented to help reinforce the definition of linear velocity. Other methods which are faster and more accurate will be explained later in this chapter.

STEP 1. Measure off linear displacement from C' to C'' [Fig. 4.3(b)]:

$$\Delta S = C'C'' = .21 \text{ in.}$$

STEP 2. The time obtained in Example 4.1 was $\Delta t = .05$ sec. Determine V_C:

$$V_C \approx V_{C_{av}} = \frac{\Delta S}{\Delta t} = \frac{.21 \text{ in.}}{.05 \text{ sec}} = 4.2 \frac{\text{in.}}{\text{sec}}$$

The velocity vector points to the right. (The exact answer is 4 in./sec.)

4.6 Linear Velocity Related to Angular Velocity

In Fig. 4.5(a), link OP rotates cw about O. Time is shown at positions OP and OP'. The average linear velocity of point P when moving from P to P' is

$$V_{P_{av}} = \frac{\Delta S}{\Delta t} = \frac{\overline{PP'}}{t_2 - t_1}$$

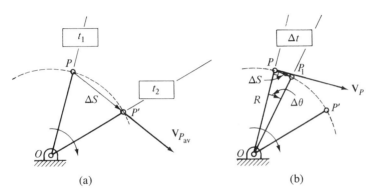

(a)   (b)

Fig. 4.5. (a) Link OP rotates about O. $V_{P_{av}} = \overline{PP'}/t_2 - t_1$. (b) Same motion as Fig. 4.5. (a); P_1 is very close to P, $V_P = \overline{OP}\omega$.

where vector $V_{P_{av}}$ is collinear to ΔS and has the same sense as ΔS. Note the length of $V_{P_{av}}$ was arbitrarily chosen.

Figure 4.5(b), which has the same motion as Fig. 4.5(a), shows an additional position OP_1 for link OP. The instantaneous linear velocity or velocity of point P when at position P can be obtained in the following manner. Let $OP =$ radius R:

$$\Delta\theta = \frac{\text{arc } PP_1}{R} \tag{1}$$

where $\Delta\theta$ is in radians. Imagine P_1 to be practically coincident with P; thus, angular displacement, linear displacement, and time are almost zero. Think of $\Delta\theta$ as $d\theta$, Δt as dt, and ΔS as $dS \approx$ arc PP_1. Substitute into Eq. (1):

$$d\theta = \frac{dS}{R} \tag{2}$$

Then

$$dS = R \, d\theta \tag{3}$$

Substitute Eq. (3) into Eq. (4.5):

$$V_P = \frac{dS}{dt} = \frac{R \, d\theta}{dt} \tag{4}$$

Recall that $d\theta/dt = \omega$. Thus

$$V_P = R\omega \tag{5}$$

The velocity of P, V_P, is a vector quantity which is drawn from point P perpendicular to position OP. The sense of the vector is the same as dS. Note that ω is the instantaneous angular velocity of link OP when link OP is at position OP.

Equation (5) is generally written as

$$V = R\omega \tag{4.6}$$

Typical units are

$$V = \text{velocity, ft/sec}$$
$$R = \text{radius, ft}$$
$$\omega = \text{angular velocity, rad/sec}$$

Equation (4.6) shows that the velocity of a point, at an instant in time, is directly proportional to the radius. That is, the longer the radius, the larger the velocity. Radius R is measured from the center of rotation.

Substituting units into Eq. (4.6), we obtain

$$\frac{\text{ft}}{\text{sec}} = \text{ft}\left(\frac{\text{rad}}{\text{sec}}\right)$$

Seemingly the units for Eq. (4.6) do not balance. The left-hand side is feet per second, and the right-hand side is foot-radians per second. Recall that the radian is not a unit but a dimensionless quantity. Thus, radians can be removed from this equation without destroying the equality.

EXAMPLE 4.4

Given: Refer to Fig. 4.6(a); link $\textcircled{L}$ rotates about the shaft center O at a constant 25 rpm cw, where $OP = 4$ in., $OQ = 2$ in., and $OR = 1$ in.

Determine: Velocity of points P, Q, and R on $\textcircled{L}$ for the position shown in Fig. 4.6(a).

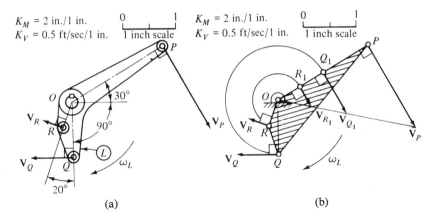

Fig. 4.6. (a) Rotating link. (b) V_Q and V_R determined graphically using similar triangles.

Solution. For this example it is not necessary to show the connecting links at points P, Q, and R. Calculate the angular velocity in radians per second:

$$\omega_L = 25 \frac{\text{rev}}{\text{min}} \left(\frac{1 \text{ min}}{60 \text{ sec}}\right)\left(\frac{2\pi \text{ rad}}{1 \text{ rev}}\right) = 2.61 \frac{\text{rad}}{\text{sec}}$$

Using Eq. (4.6), determine the velocity for points P, Q, and R:

$$V_P = \overline{OP}\omega_L = 4 \text{ in.} \left(\frac{1 \text{ ft}}{12 \text{ in.}}\right) 2.61 \frac{\text{rad}}{\text{sec}} = .870 \frac{\text{ft}}{\text{sec}}$$

$$V_Q = \overline{OQ}\omega_L = \frac{2}{12}(2.61) = .435 \frac{\text{ft}}{\text{sec}}$$

$$V_R = \overline{OR}\omega_L = \frac{1}{12}(2.61) = .218 \frac{\text{ft}}{\text{sec}}$$

Velocities are represented in Fig. 4.6(a). Two scales are required: The mechanism scale $K_M = 2 \text{ in.}/1 \text{ in.}$, and the velocity scale $K_V = .5 \text{ ft/sec}/1 \text{ in.}$ Convert velocities to drawing dimensions:

$$V_P = V_P\left(\frac{1}{K_V}\right) = .870 \frac{\text{ft}}{\text{sec}} \left(\frac{1 \text{ in.}}{.5 \text{ ft/sec}}\right) = 1.740 \text{ in.}$$

$$V_Q = .435 \left(\frac{1}{.5}\right) = .870 \text{ in.}$$

$$V_R = .218 \left(\frac{1}{.5}\right) = .435 \text{ in.}$$

Draw V_P, V_Q, and V_R perpendicular to the appropriate radius line. The sense of the velocity vectors is consistent with the direction of ω_L.

In Example 4.4, V_Q and V_R can be determined graphically by similar triangles. The justification for this construction is as follows. Using Eq. (4.6),

$$V_P = \overline{OP}\omega_L$$

Therefore

$$\omega_L = \frac{V_P}{OP} \tag{1}$$

Also,

$$V_Q = \overline{OQ}\omega_L$$

Therefore

$$\omega_L = \frac{V_Q}{OQ} \tag{2}$$

Equate Eqs. (1) and (2):

$$\frac{V_P}{OP} = \frac{V_Q}{OQ} \tag{3}$$

This is a proportion which can be graphically constructed as similar triangles. Equation (3) is true whether ω_L is constant or varies with time.

To illustrate how velocities are determined by similar triangles, Example 4.4 will be solved using this graphical method. The completed construction is shown in Fig. 4.6(b). The procedure is outlined in step form as follows:

STEP 1. Calculate V_P. Now lay out V_P to scale [Fig. 4.6(b)].

STEP 2. Measure off radii OQ and OR along OP; this locates Q_1 and R_1. An easier method is to use point O as center; swing an arc through Q which intersects OP at Q_1. Do the same for point R.

STEP 3. Construct a line from O to the tip of V_P.

STEP 4. Draw a line from Q_1 to the line previously constructed (step 3) which is either parallel to V_P or perpendicular to OP. Use the same procedure for R_1. Similar triangles are now constructed. Note that

$$\frac{V_P}{OP} = \frac{V_{Q_1}}{OQ_1} = \frac{V_{R_1}}{OR_1}$$

STEP 5. Draw a velocity line or line of action of V_Q at Q. Do the same at R.

STEP 6. Using dividers, mark off the magnitude of V_{Q_1} onto the line of action of V_Q. Use the same procedure for V_{R_1}. Draw in arrowheads on V_Q and V_R. This completes the construction.

4.7 Velocity Components

A relatively fast and simple technique used in velocity analysis is the velocity component method, also called the resolution or effective component method. This method is based on the kinematic assumption that a link is rigid.

In Fig. 4.7(a), link L is shown moving in translation. If V_A is greater than V_B, the link will be crushed. On the other hand, if V_B is greater than V_A, the link will be pulled apart. Keeping the link intact requires that the velocity along line AB be the same; that is, $V_A = V_B$.

Figure 4.7(b) shows a more general situation; here link L rotates cw about point O, and V_A and V_B are known. The instantaneous velocities V_A and V_B are resolved into components parallel and perpendicular to AB. Thus

$$V_A = {}_{AB}V_A + {}_{\perp AB}V_A \quad \text{and} \quad V_B = {}_{AB}V_B + {}_{\perp AB}V_B$$

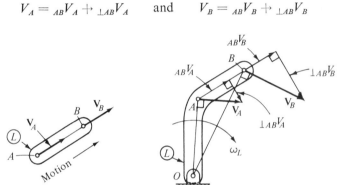

(a) (b)

Fig. 4.7. (a) L moves in translation, $V_A = V_B$. (b) L rotates, ${}_{AB}V_A = {}_{AB}V_B$.

Where ${}_{AB}V_A$ is read as velocity of A along line AB and ${}_{\perp AB}V_A$ is read as velocity of A perpendicular to line AB; ${}_{AB}V_B$ and ${}_{\perp AB}V_B$ are read in a similar manner. To maintain a constant distance between points A and B it is required that the components along line AB be equal, ${}_{AB}V_A = {}_{AB}V_B$. Note that in this illustration, $V_A \neq V_B$.

At an instant in time, any two points on a link must have equal velocity components along the line connecting these points. The component value may be the velocity itself, as in Fig. 4.7(a). When constructing components there is sometimes confusion as to where the right angle should be. A simple rule to remember is that the right angle is always drawn on the component head. *Component* refers to the velocity component acting along the line connecting the points.

The following example illustrates how the component method is applied to a mechanism.

EXAMPLE 4.5

Given: Figure 4.8(a); crank AB, of an offset slider crank ABC, rotates at 2000 rpm cw. $AB = \frac{7}{16}$ in., $BC = 1\frac{9}{32}$ in., and the offset $= \frac{9}{32}$ in.

Determine: Blade velocity (feet per second) for the linkage position shown in Fig. 4.8(a), section X-X. Use the component method.

Solution. This mechanism uses an offset slider crank to obtain a quick return motion for the blade. An electric motor transmits motion to the crank through a pair of helical gears.

Observe that point C, on the reciprocating plunger, has the same velocity as the blade. Thus the blade velocity will be determined by calculating the velocity for point C. Both the blade and plunger are rigidly fastened together, and both members are translating. The completed solution is shown in Fig. 4.8(b).

STEP 1. Calculate V_B using Eq. (4.6), $AB = \frac{7}{16}$ in. $= .438$ in.:

$$V_B = \overline{AB}\omega_{AB} = \frac{.438}{12}\left[\frac{2000(2\pi)}{60}\right] = 7.63\,\frac{\text{ft}}{\text{sec}}$$

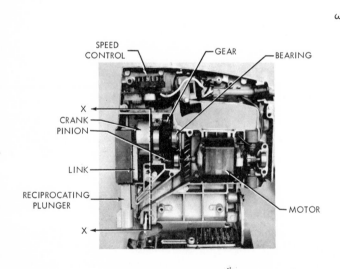

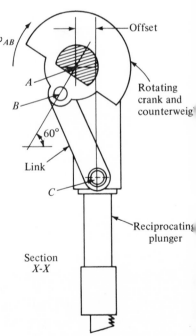

Fig. 4.8. (a) Cutaway view of a portable saber saw. Section X-X *shows offset slider crank* ABC. (Courtesy of Montgomery Ward.)

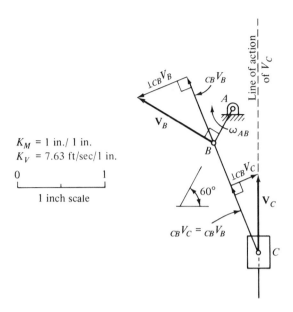

$K_M = 1$ in./ 1 in.
$K_V = 7.63$ ft/sec/1 in.

1 inch scale

Fig. 4.8. (Cont.) (b) V_C determined by the component method.

STEP 2. Assume a velocity scale; let $K_V = 7.63$ ft/sec/1 in.

STEP 3. Draw V_B (scaled) perpendicular to AB at B.

STEP 4. At B, construct a right triangle so that $V_B = {}_{CB}V_B + {}_{\perp CB}V_B$.

STEP 5. Using dividers or a ruler, transfer ${}_{CB}V_B$ to point C. Label vector ${}_{CB}V_C$.

STEP 6. At C, construct a right triangle so that $V_C = {}_{CB}V_C + {}_{\perp CB}V_C$. Construct V_C on the line of action of V_C.

STEP 7. Scale off V_C; $V_C = \frac{7}{8}$ in. Thus

$$V_C = \left(\frac{7}{8} \text{ in.}\right)\left(\frac{7.63 \text{ ft/sec}}{1 \text{ in.}}\right) = 6.68 \frac{\text{ft}}{\text{sec}}$$

The blade velocity $= V_C = 6.68$ ft/sec upward.

Since the perpendicular components are not directly used in the solution, we can save time by not labeling these vectors.

4.8 Instant Center of Rotation

It was stated in Chapter 3 that a link moving with plane motion moves with either translation, rotation, or combination. Motion occurred over a relatively long span of time, for example, 4 sec, 2 min, 3 hr. This section is concerned with plane motion at an instant in time.

At an instant in time a link moving with plane motion can be regarded as instantaneously rotating about a point or axis called the instant center of rotation or simply the instant center. Rotation is always relative to another link. In this text the instantaneous rotation of a link will be taken relative to the earth. Linear and angular velocities related to the instant center will be absolute quantities. Other names for the instant center are instantaneous center and centro.

Consider link T shown in Fig. 4.9(a). Link T clearly rotates about point O. Thus the instant center of rotation of link T, relative to earth E, is at point O. The instant center O in this figure happens to be a permanent center of rotation. That is, link T permanently rotates about point O. If T were rotated to another position, the instant center would still be at connection O. The instant center of a "floating link" usually changes position from instant to instant. A floating link is not connected to the frame or earth, as is the connecting rod of a slider crank.

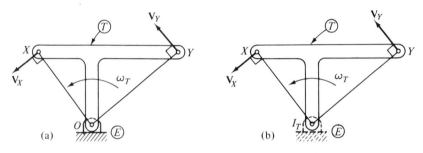

Fig. 4.9. (a) Instant center (permanent center) of T is at O. (b) Instant center of T is at I_T.

Also shown in Fig. 4.9(a) are the absolute velocities of points X and Y. V_X and V_Y are drawn perpendicular to their respective radius line. A radius line, by definition, passes through the center of rotation. Note that the velocity of the instant center or permanent center O is zero, $V_O = 0$. In this text the instant center will always have a zero velocity.

Keep in mind the similarities between Figs. 4.9(a) and 4.9(b) while reading this paragraph. Figure 4.9(b) shows one link of a moving mechanism (mechanism not shown). V_X and V_Y are completely known absolute velocities. To locate the instant center rotation for link T requires the construction of lines perpendicular to V_X and V_Y at points X and Y. These constructed lines are the radius lines. The intersection of these lines locates the instant center. Label this point I_T or I_{XY}. I_T need not be physically on T. I_T can be thought of as the location of an imaginary pin connection between T and E. To help visualize this connection, a dotted ground symbol is shown at I_T. This imaginary connection lasts only for an instant of time. Again it is emphasized that the absolute velocity of the instant center is zero, $V_{I_T} = 0$.

In Fig. 4.9(a), point O is a permanent center of rotation, whereas in Fig. 4.9(b), not enough information was given to determine if point I_T is a permanent or instant center.

4.9 Locating the Instant Center of a Link

There are several methods available for locating the instant center. One method is to study the motion of the link. This method was somewhat implied in Fig. 4.3(b). The method which will be used in this text is essentially a velocity method. The following examples illustrate how the instant center is located. All the mechanisms discussed are moving.

Figures 4.10(a) and 4.10(b) show a four-bar linkage $ABCD$. Points A and

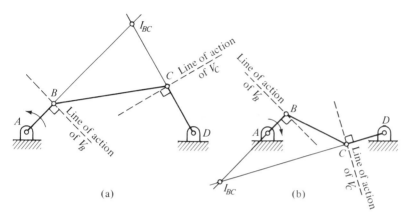

Fig. 4.10. *Instant center* I_{BC} *for a four-bar linkage* ABCD.

D are permanent centers of rotation for links AB and CD, respectively. V_B and V_C are unknown; however, the line of actions for V_B and V_C can be established. V_B is perpendicular to AB at B, and V_C is perpendicular to CD at C. The instant center for link BC, I_{BC}, is located as follows. Draw a line perpendicular to the line of action of V_B at B; do the same for V_C at C. In other words, simply extend lines AB and CD. The intersection of these lines locates I_{BC}.

Refer to slider crank ABC shown in Figs. 4.11(a) and 4.11(b). V_B and V_C are unknown. The line of action of V_B is perpendicular to AB point B. The path of point C is parallel to the slider guide; therefore the line of action of V_C is drawn through C parallel to the slider guide. To locate the instant center for connecting rod BC, I_{BC}, draw a line perpendicular to V_B at B; do the same for V_C at C. The intersection of these lines locates I_{BC}. In Fig. 4.11(b), I_{BC} is at point C; thus the velocity of I_{BC} or C is zero at this instant.

In Fig. 4.12(a), gear ⓖ is driven by a rack gear ⓜ; rack ⓡ is stationary. It can be easily demonstrated that the linear displacement for the moving rack ⓜ is twice that for gear center O. Since both displacements occur in the same span of time, it can be concluded that the linear velocity of ⓜ is twice that for point O. For the instant shown in Fig. 4.12(a), the velocity of rack ⓜ is the same as contact point Y on gear ⓖ. To locate the instant center for gear ⓖ, I_G, draw a line perpendicular to V_Y at Y; do the same for V_O at O. Unfortunately this

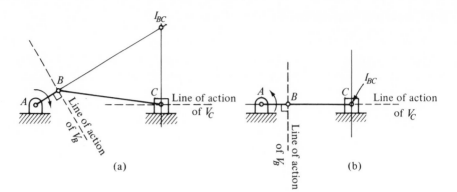

Fig. 4.11. *Instant center I_{BC} for a slider crank ABC.*

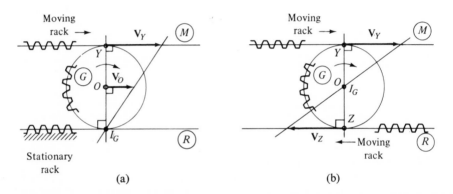

Fig. 4.12. *Instant center I_G for a gear.*

does not locate I_G because both drawn lines are collinear. More construction is required. Draw a line which connects the heads of V_Y and V_O. Intersection of this line with the previously drawn line locates I_G. I_G is located at the point of contact with the stationary rack. V_{I_G} is zero, and the velocities of points on Ⓖ are proportional to the distance from I_G.

Figure 4.12(b) shows a gear Ⓖ being rotated by two rack gears Ⓜ and Ⓡ. Both racks have the same magnitude of velocity. Points Y and Z are located on Ⓖ, where V_Y and V_Z are equal in magnitude and opposite in sense. To locate I_G, draw a line perpendicular to V_Y at Y (or V_Z at Z); also, draw a line which connects vector heads V_Y and V_Z. The intersection of these lines locates I_G. I_G is located at the gear center. The velocity of I_G or point O is zero.

4.10 Determining Velocity
by Instant Center

The previous sections answered such questions as what is the instant center and how is it located. We are now ready to use the instant center to determine velocity.

Figure 4.13 (a) shows a four-bar linkage $WXYZ$. It is desirable to determine

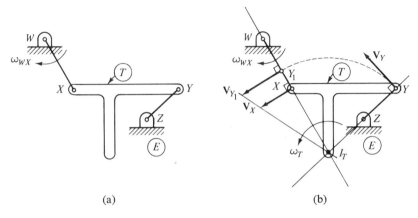

(a)   (b)

Fig. 4.13. (a) Four-bar linkage. (b) V_Y determined graphically using similar triangles.

the linear velocity of point Y and the angular velocity of coupler link T. Angular velocity ω_{WX} is given. The solution is outlined as follows. Calculate V_X, $V_X = WX\omega_{WX}$. V_X is laid out to scale at point X; refer to Fig. 4.13(b). To locate I_T, extend lines WX and YZ; the intersection locates I_T. Since link T rotates about I_T at this instant, the linear velocity of points on T (or imagined to be on T) are proportional to the distance from I_T. Using point I_T as a pivot point, swing an arc through Y which intersects line WXI_T. Label this point Y_1. Construct similar triangles to determine V_{Y_1}. The magnitude of $V_{Y_1} = V_Y$. Transfer the magnitude of V_{Y_1} to point Y. Construct V_Y perpendicular to ZY at Y. The angular velocity ω_T can be determined from $V = R\omega$. Solving for ω_T,

$$\omega_T = \frac{V_X}{XI_T} \text{ ccw} \tag{1}$$

or

$$\omega_T = \frac{V_Y}{YI_T} \text{ ccw} \tag{2}$$

Equation (1) may be a little more accurate than Eq. (2) because V_Y was determined graphically in Eq. (2). If desirable, the linear velocity of Y can be calculated by equating Eqs. (1) and (2); thus

$$\frac{V_X}{XI_T} = \frac{V_Y}{YI_T}$$

Solving for V_Y,

$$V_Y = V_X \left(\frac{YI_T}{XI_T}\right) \quad \text{or} \quad V_Y = (WX\omega_{WX})\left(\frac{YI_T}{XI_T}\right)$$

EXAMPLE 4.6

Given: Figure 4.14(a); four-bar linkage $ABCD$. AB rotates at 600 rad/min cw. $AB = 2$ in., $BC = 4$ in., $CD = 3$ in., $AD = 4$ in., $BCE = 6$ in., and $EP = 1$ in.

Determine: V_P and ω_{EP} for the position shown. Use the instant center method.

Solution

STEP 1. Refer to Fig. 4.14(b). Calculate V_B:

$$V_B = AB\omega_{AB} = \left(\frac{2}{12}\right)600 = 100\frac{\text{ft}}{\text{min}}$$

STEP 2. Assume $K_V = 100$ ft/min/1.5 in. Draw V_B (to scale) perpendicular to AB at B.

STEP 3. Locate I_{BC}. Extend lines BA and CD. The intersection locates I_{BC}.

STEP 4. Using I_{BC} as the center of rotation, swing an arc through P which intersects extended line AB, label this point P_1.

STEP 5. Similar triangles are constructed to determine V_{P_1}:

$$V_{P_1} = .74 \text{ in.} \left(\frac{100 \text{ ft/min}}{1.5 \text{ in.}}\right) = 49.4\frac{\text{ft}}{\text{min}}$$

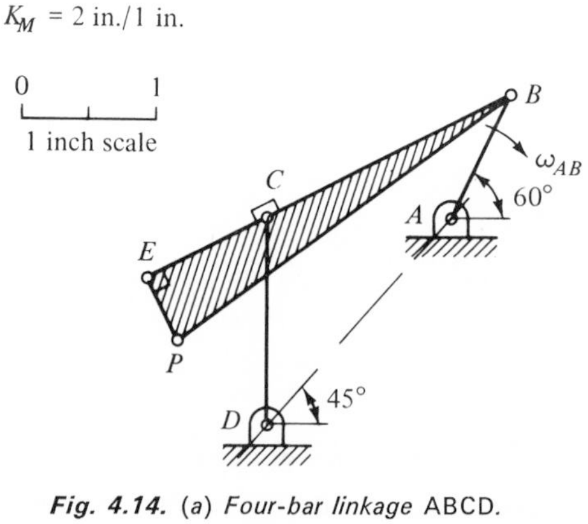

$K_M = 2$ in./1 in.

0   1
1 inch scale

Fig. 4.14. (a) *Four-bar linkage* ABCD.

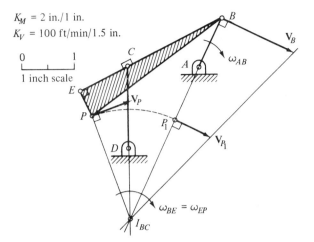

Fig. 4.14. *(Cont.) (b)* V_P *determined by similar triangles. The angular velocity of EP is* $\omega_{EP} = V_B/BI_{BC}$.

STEP 6. The magnitude of $V_P = V_{P_1}$; therefore, $V_P = 49.4\ \text{ft/min}$. Mark off V_P at P where V_P is perpendicular to PI_{BC}.

STEP 7. Determine ω_{EP}:

$$BI_{BC} = 4.1\ \text{in.}\left(\frac{2\ \text{in.}}{1\ \text{in.}}\right) = 8.2\ \text{in.}$$

$$\omega_{EP} = \omega_{BE} = \frac{V_B}{BI_{BC}} = \frac{100\ \text{ft/min}}{(8.2/12)\ \text{ft}} = 146.3\,\frac{\text{rad}}{\text{min}}\,\text{cw}$$

or

$$\omega_{EP} = \frac{V_P}{PI_{BC}} = \frac{49.4\ \text{ft/min}}{(4/12)\ \text{ft}} = 148.3\,\frac{\text{rad}}{\text{min}}\,\text{cw}$$

Both answers to ω_{EP} are in close agreement.

EXAMPLE 4.7

Given: The variable displacement pump shown in Fig. 4.15(a). The drive shaft rotates at 1800 rpm cw. $OA = .5\ \text{in.}$, the pitch diameter for the satellite pinion $\textcircled{P}$ is 1 in., the pitch diameter for the ring gear is 2 in., and $BC = 3\ \text{in.}$

Determine: For the position shown, the piston velocity and angular velocity for the satellite gear $\textcircled{P}$.

Solution. The drive shaft, crank plate, and offset crank pin rotate together at 1800 rpm cw. The crank pin moves the satellite pinion, and the pinion revolves ccw on the internal teeth of the ring gear. Consider the ring gear stationary. The pinion plate is rigidly fastened to the pinion. One end of the connecting rod is attached to the outer rim of the pinion plate. As the pinion plate revolves ccw, point B translates, moving the connecting rod and piston.

When the ring gear diameter is two times the pinion diameter, as in this mechanism,

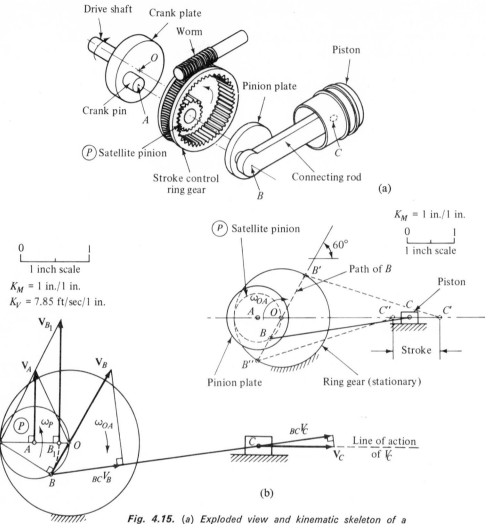

Fig. 4.15. (a) *Exploded view and kinematic skeleton of a variable displacement pump.* (b) *Solution to pump mechanism.* (Courtesy of Design News, *Czechoslovak Patent 112,865.*)

point B traces a straight line. Point B lies at the pitch circumference of the satellite pinion. The end positions for the path of B (B' and B'') determine the piston stroke. Rotation of the ring gear, by means of a worm and worm gear, changes the path of B, thus changing the piston stroke.

STEP 1. Calculate V_A:

$$V_A = OA\omega_{OA} = \frac{.5}{12}\left[1800\left(\frac{2\pi}{60}\right)\right] = 7.85 \frac{ft}{sec}$$

STEP 2. Assume that $K_V = 7.85$ ft/sec/1 in. See Fig. 4.15(b). Draw V_A (to scale) perpendicular to OA at A.

STEP 3. Locate I_P. Gear ⓟ rolls without slipping on the stationary ring gear. I_P is at the point of contact.

STEP 4. Using I_P as the center of rotation, swing an arc through B which intersects the extended line I_PA. Label this point B_1.

STEP 5. Construct similar triangles and determine V_{B_1}.

STEP 6. The magnitude of $V_B = V_{B_1}$. Mark off V_B at point B.

STEP 7. Determine V_C by the component method; $_{BC}V_B = {_{BC}}V_C$. The piston velocity is

$$V_C = 1.08 \text{ in.} \left(\frac{7.85 \text{ ft/sec}}{1 \text{ in.}}\right) = 8.48 \frac{\text{ft}}{\text{sec}}$$

STEP 8. Determine ω_P, where $V_A = 7.85$ ft/sec and $V_B = 13.20$ ft/sec:

$$\omega_P = \frac{V_A}{AI_P} = \frac{7.85 \text{ ft/sec}}{(.5/12) \text{ ft}} = 188 \frac{\text{rad}}{\text{sec}} \text{ ccw}$$

or

$$\omega_P = \frac{V_B}{BI_P} = \frac{13.20 \text{ ft/sec}}{(.86/12) \text{ ft}} = 188 \frac{\text{rad}}{\text{sec}} \text{ ccw}$$

4.11 Relative Linear Velocity

Absolute linear velocity, or simply velocity, is the velocity of a point relative to a stationary point on earth. Relative linear velocity is with respect to a moving point.

Figure 4.16(a) illustrates an example of relative velocity, where the airplane, helicopter, and ground observer will be treated as points. The velocity of the

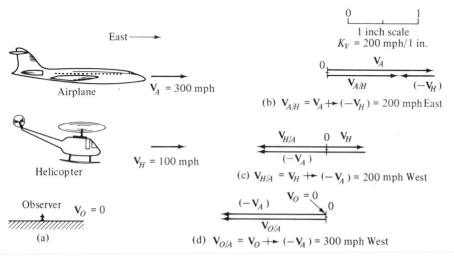

Fig. 4.16. Absolute and relative velocity.

helicopter relative to the earth, or to the ground observer O, is 100 mph east. Symbolically this is $V_{H/E} = V_{H/O} = V_H = 100$ mph east. The airplane velocity relative to the earth is $V_{A/E} = V_A = 300$ mph east. Note that relative to the helicopter pilot the airplane velocity is 200 mph east; $V_{A/H} = 200$ mph east. This is determined either from observation or from the relative vector equation

$$V_{A/H} = V_A \rightarrow V_H \quad \text{or} \quad V_{A/H} = V_A \,{+}\!\!\!+\, (-V_H) \tag{4.7}$$

The graphical solution of Eq. (4.7) is shown in Fig. 4.16(b). Since the helicopter and airplane are translating parallel to each other, Eq. (4.7) could have been solved algebraically rather than vectorially.

The airplane pilot sees the helicopter moving at 200 mph west; $V_{H/A} = 200$ mph west. The vector equation is

$$V_{H/A} = V_H \rightarrow V_A = V_H \,{+}\!\!\!+\, (-V_A)$$

The solution to this equation is shown in Fig. 4.16(c).

Comparison of Figs. 4.16(b) and 4.16(c) shows that

$$V_{A/H} = -V_{H/A} \quad \text{or} \quad -V_{A/H} = V_{H/A} \tag{4.8}$$

Note that the airplane pilot sees the ground observer move at a velocity of 300 mph west; refer to Fig. 4.16(d).

4.12 Pure Rolling

If two rigid bodies are in pure rolling, no slip, the velocity of their contact points are equal. Another way of saying this is that the relative velocity between contact points is zero.

Figure 4.17 shows two spur gears, which can be considered as two disks, in pure rolling. Actually there is sliding between gear teeth; however, at the contact point or pitch point no sliding occurs. Contact points are designated X and Y—X on $\textcircled{D}$ and Y on $\textcircled{F}$. Gear $\textcircled{D}$ is the driver, and $\textcircled{F}$ is the follower:

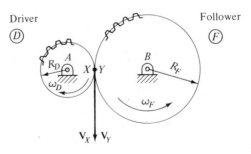

Fig. 4.17. Gears considered to be in pure rolling.

$$V_X = R_D \omega_D \tag{1}$$

$$V_Y = R_F \omega_F \tag{2}$$

There is no slip between $\textcircled{D}$ and $\textcircled{F}$:

$$V_X = V_Y \tag{3}$$

Substitute Eqs. (1) and (2) into Eq. (3):

$$R_D \omega_D = R_F \omega_F \tag{4}$$

Rearranging,

$$\frac{\omega_D}{\omega_F} = \frac{R_F}{R_D} \tag{4.9}$$

Equation (4.9) shows that the angular velocity ratio is inverse to the radius ratio. Comparatively speaking, a small gear has a large angular velocity and a large gear has a small angular velocity.

The relative velocity between X and Y can be shown to be zero:

$$V_{X/Y} = V_X + (-V_Y)$$

Since $V_X = V_Y$, $V_{X/Y} = 0$.

EXAMPLE 4.8

Given: Variable-speed traction drive; refer to Fig. 4.18. The input shaft rotates at 1500 rpm cw (viewed from the right side).

Determine: Output speed; assume no slip between double-cone roller and disks.

Solution. The driving disk imparts motion to a driven disk through a double-cone roller. The output speed is adjusted by rotating the leadscrew, which shifts the roller vertically. Speed changes are made while the unit is in motion. To reduce slipping, pressure on the roller is maintained by a load-sensitive output shaft mechanism. If load on the output shaft increases, pressure on the roller increases.

STEP 1. Refer to Fig. 4.18. Label the contact points: X on the driving disk, Y and Z on the roller, and P on the driven disk.

STEP 2. Calculate V_X:

$$V_X = OX \omega_{OX} = \frac{2}{12}\left[\frac{1500(2\pi)}{60}\right] = 8.73 \text{ ft/sec}$$

STEP 3. Assume that $K_V = 8.73$ ft/sec/1 in. Draw in V_X.

STEP 4. No slip; $V_Y = V_X$.

STEP 5. Both Y and Z are an equal distance from the axis of rotation. Thus

$$|V_Y| = |V_Z|$$

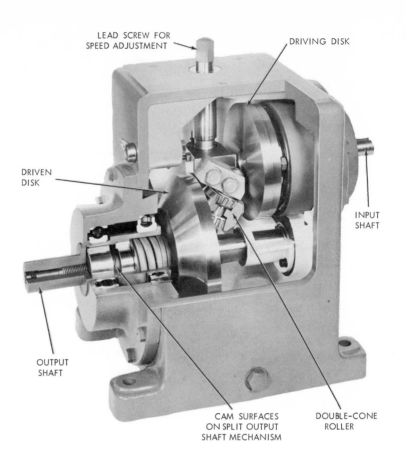

LEAD SCREW FOR SPEED ADJUSTMENT

DRIVING DISK

DRIVEN DISK

INPUT SHAFT

OUTPUT SHAFT

CAM SURFACES ON SPLIT OUTPUT SHAFT MECHANISM

DOUBLE-CONE ROLLER

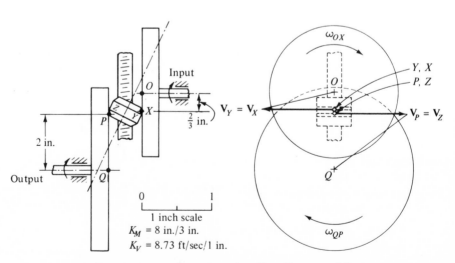

Input

$\mathbf{V}_Y = \mathbf{V}_X$

$\mathbf{V}_P = \mathbf{V}_Z$

ω_{OX}

ω_{QP}

Y, X

P, Z

$\frac{2}{3}$ in.

2 in.

Output

0   1

1 inch scale

K_M = 8 in./3 in.

K_V = 8.73 ft/sec/1 in.

Fig. 4.18. *Variable-speed traction drive.* (Courtesy of Cone-Trol Unicum.)

116

Draw in V_Z.

STEP 6. No slip; $V_P = V_Z$.

STEP 7. Determine output speed:

$$\omega_{QP} = \frac{V_P}{QP} = \frac{8.73}{(\frac{2}{12})}\left(\frac{60}{2\pi}\right) = 500 \text{ rpm cw}$$

This type of problem can be solved without converting units. An alternative solution follows:

$$V_X = OX\omega_{OX} \tag{1}$$

$$V_P = QP\omega_{QP} \tag{2}$$

and

$$|V_P| = |V_X| \tag{3}$$

Substitute Eqs. (1) and (2) into Eq. (3):

$$QP\omega_{QP} = OX\omega_{OX}$$

Rearranging,

$$\omega_{QP} = \frac{OX}{QP}\omega_{OX} \tag{4}$$

Substitute numerical values into Eq. (4):

$$\omega_{QP} = \frac{\frac{2}{3}}{2}(1500) = 500 \text{ rpm cw}$$

EXAMPLE 4.9

Given: Figure 4.19(a); an eccentric cam rotates at 100 rpm ccw. Roller follower ℝ is pinned to BC at B. Eccentricity $= 1$ in., $\overline{BC} = 3$ in., the cam diameter $= 3$ in., and the roller diameter $= 1$ in.

Determine: ω_{BC}; assume no slipping between cam and roller.

Solution

STEP 1. Refer to Fig. 4.19(b). To locate contact points X and Y, draw line OB. Label X on ⒟, and Y on ℝ.

STEP 2. Calculate V_X, where $AX = 2.05$ in.:

$$V_X = AX\omega_D = \left[\frac{2.05}{12}\right]\left[\frac{100(2\pi)}{60}\right] = 1.79 \frac{\text{ft}}{\text{sec}}$$

STEP 3. Assume that $K_V = 1.79$ ft/sec/.75 in. Draw V_X (to scale) perpendicular to AX at X.

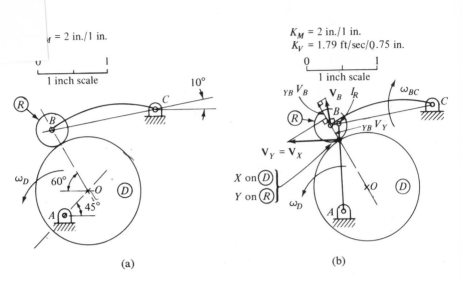

(a)   (b)

Fig. 4.19. (a) *Cam and oscillating roller follower. (b) Assume no slip;* $V_Y = V_X$ *and* $\omega_{BC} = V_B/BC$.

STEP 4. No slip; $V_Y = V_X$.
STEP 5. Determine V_B by the component method; $_{YB}V_Y = _{YB}V_B$:

$$V_B = .38 \text{ in.} \left(\frac{1.79}{.75}\right) = .91 \frac{\text{ft}}{\text{sec}}$$

STEP 6. Determine ω_{BC}:

$$\omega_{BC} = \frac{.91 \text{ ft/sec}}{\frac{3}{12} \text{ ft}} = 3.63 \frac{\text{rad}}{\text{sec}} \text{ cw}$$

It is interesting to note that if sliding existed between the cam and the roller, V_B would still be .91 ft/sec. By assuming pure rolling, the instant center of $\circledR$ is located to the right of B. Specifically, I_R is located by drawing lines perpendicular to V_Y and V_B through points Y and B. Roller $\circledR$ is rotating cw about I_R at this instant. The angular velocity of the roller can be calculated from $\omega_R = V_B/BI_R = V_Y/YI_R$.

4.13 Sliding

Two rigid bodies which are in contact and are not in pure rolling are said to be sliding. The exception to this is when two contacting bodies are stationary or move as one rigid body.

Sliding can best be described by a simple example. Figure 4.20 shows two books in contact; $\circledB$ is stationary and $\circledA$ slides on surface $\circledB$. The surface or

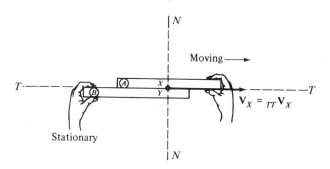

Fig. 4.20. Ⓐ *slides on the stationary surface of* Ⓑ.

line on which sliding takes place is labeled TT and is referred to as the tangent line. Normal line NN is perpendicular to TT. The intersection of NN and TT locates contact points X and Y, where X is on Ⓐ and Y is on Ⓑ. The proper analysis of sliding requires that the velocity of X and Y be resolved into components along NN and TT. The velocity components of Y are symbolized as $_{NN}V_Y$ and $_{TT}V_Y$. Read $_{NN}V_Y$ as the velocity of Y along normal line NN or simply the normal velocity of Y. Read $_{TT}V_Y$ as the velocity of Y along tangent line TT or simply the tangent velocity of Y. The vector equation for V_Y is

$$V_Y = {}_{NN}V_Y + {}_{TT}V_Y \tag{1}$$

Since body Ⓑ or point Y is stationary, it follows that $V_Y = 0$, $_{NN}V_Y = 0$, and $_{TT}V_Y = 0$.

The velocity equation for X is

$$V_X = {}_{NN}V_X + {}_{TT}V_X \tag{2}$$

Body Ⓐ or point X moves only in the TT direction. Thus

$$V_X = {}_{TT}V_X \qquad \text{and} \qquad {}_{NN}V_X = 0$$

Comparison of the results from Eqs. (1) and (2) shows that $V_Y \neq V_X$; however, $_{NN}V_Y = {}_{NN}V_X$. It is not coincidental that $_{NN}V_Y = {}_{NN}V_X$. If $_{NN}V_Y$ were not equal to $_{NN}V_X$, X would move either toward Y or away from Y along NN. Theoretically this is not possible. If X moves toward Y, a crushing condition exists, which violates the rigid body assumption. If X moves away from Y (other than along TT), both bodies separate, which is contrary to the stated condition of being in contact.

Figure 4.21(a) shows an example of sliding where both bodies or books are in motion. Sliding takes place along line TT. The motion of each book can be separated into two distinct motions:

Motion 1. Book Ⓐ moves to the right along TT and Ⓑ to the left along TT.

Motion 2. Both books move together along path 1.

Figure 4.21(b) shows one position of the books. Note that $V_Y \neq V_X$; however, because of contact between the books, the velocity component of X and Y normal to their respective book are equal; that is, $_{NN}V_Y = {}_{NN}V_X$.

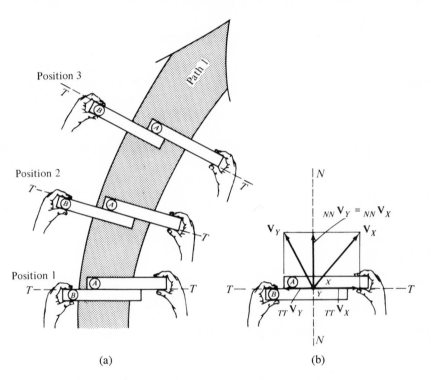

(a)   (b)

Fig. 4.21. (a) Sliding; Ⓐ and Ⓑ are in motion. (b) Position 1 of Fig. 4.21. (a). $V_Y \neq V_X$; however, $_{NN}V_Y = {}_{NN}V_X$.

Figure 4.22 shows the case where Ⓐ and Ⓑ move together as one rigid body. This is neither sliding nor pure rolling. It is evident that $V_Y = V_X$.

A velocity analysis of a sliding mechanism sometimes requires a simplification of the kinematic skeleton. For example, Fig. 4.23(a) shows a slider Ⓢ which moves in a slotted link Ⓛ. The slider is pinned to link AB at B. For purposes of analysis the slider Ⓢ is reduced in size to a point; see Fig. 4.23(b). Thus it can be said that slider Ⓢ and point B are one point. To be consistent with the previous discussion, the contact points at B are labeled X and Y, where X is on Ⓚ and Y on Ⓛ. It should be realized that the motion of Ⓛ was not altered by

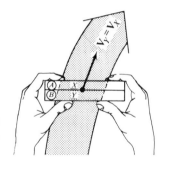

Fig. 4.22. Ⓐ and Ⓑ move together as one rigid body, $V_Y = V_X$.

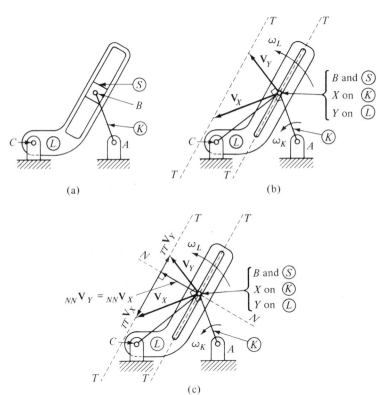

Fig. 4.23. (a) Sliding mechanism. (b) Velocity analysis. (c) Velocity analysis showing components.

this simplification. Let us now analyze a typical problem concerning this mechanism; given ω_K, determine ω_L. The solution is outlined as follows; refer to Fig. 4.23(b).

STEP 1. Calculate V_X, where

$$V_X = AX\omega_K$$

Lay out V_X to scale.

STEP 2. Sliding takes place along the slot; label this line TT. Construct another line TT on the head of vector V_X, which is parallel to the first TT line.

STEP 3. Determine V_Y. V_Y is drawn perpendicular to $\overline{YC}$ at Y. The vector extends from Y to line TT.

STEP 4. Determine ω_L:

$$\omega_L = \frac{V_Y}{\overline{YC}} \text{ ccw}$$

Step 3 is a crucial step. The justification for step 3 is that normal velocity components of the contact points are equal $(_{NN}V_X = {}_{NN}V_Y)$. Normal components are explicitly shown in Fig. 4.23(c), where

$$V_X = {}_{NN}V_X + {}_{TT}V_X$$
$$V_Y = {}_{NN}V_Y + {}_{TT}V_Y$$

and

$$_{NN}V_X = {}_{NN}V_Y$$

Usually a solution does not require all the details shown in Fig. 4.23(c). Thus the preferred solution is Fig. 4.23(b).

EXAMPLE 4.10

Given: Figure 4.24(a); cam © rotates at 15 rpm about a knitting machine center. Cam angle is 45 deg.

Determine: Needle velocity and the sliding velocity between the needle and the cam.

Solution. Cam © is part of a machine section (see the photograph) which is fastened onto a ring. This ring has a radius of 15 in. and rotates at 15 rpm. As the cam moves from left to right, it pushes the needle upward. The needle is set between stationary guides and is held against the machine wall by a circumferential spring.

STEP 1. Refer to Fig. 4.24(b). Sliding takes place along the cam surface; label this TT. NN is perpendicular to TT at the point of contact.

STEP 2. Label the contact points—X on © and Y on Ⓝ.

STEP 3. Calculate V_X:

$$V_X = R\omega_C = \frac{15}{12}\left[\frac{15(2\pi)}{60}\right] = 1.96 \frac{\text{ft}}{\text{sec}}$$

STEP 4. Let $K_V = 1.96$ ft/sec/1 in. Lay out V_X.

STEP 5. Determine V_Y. Construct a line TT parallel to the original TT line which passes through the head of vector V_X. V_Y extends vertically from Y to line TT. Since the needle is translating, both the needle and point Y have the same velocity. The needle velocity $= V_Y = 1.96$ ft/sec (V_Y was scaled from the drawing).

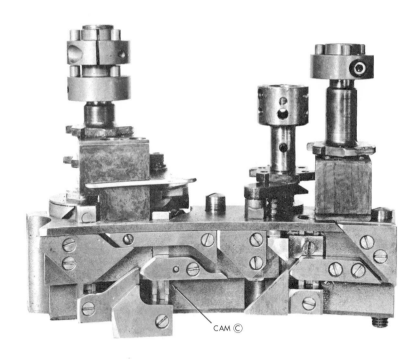

CAM Ⓒ

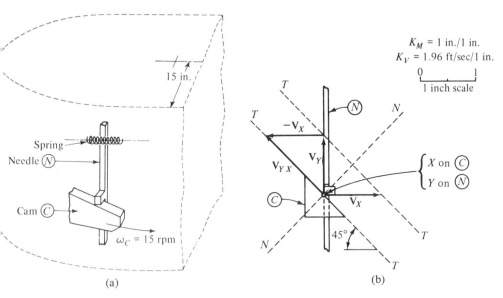

15 in.

Spring
Needle Ⓝ

Cam Ⓒ

ω_C = 15 rpm

(a)

K_M = 1 in./1 in.
K_V = 1.96 ft/sec/1 in.

0 _____ 1
1 inch scale

T

T

$-\mathbf{V}_X$

$\mathbf{V}_{YX}$

$\mathbf{V}_Y$

Ⓝ

N

$\begin{cases} X \text{ on } Ⓒ \\ Y \text{ on } Ⓝ \end{cases}$

Ⓒ

$\mathbf{V}_X$

N

45°

T

T

(b)

Fig. 4.24. (a) Rear view of a knitting machine section and a simplified pictorial of a circular knitting machine. The machine section is bolted onto a 30-in. diameter ring with two bolts (see the photograph). (b) Solution to cam problem. (Courtesy of Wildman Jacquard Co.)

123

STEP 6. Sliding velocity is defined as the relative velocity between Y and X. Therefore

$$V_{Y/X} = V_Y \to V_X = V_Y +\!\!\!\!\!\to (-V_X)$$
$$V_{Y/X} = 2.77 \text{ ft/sec}$$

EXAMPLE 4.11

Given: Figure 4.25(a); rotating slider crank mechanism. Input into gear G is 1000 rpm ccw (viewed from the right side). The dimensions are $AB = .5$ in., $BC = 2$ in., and $CD = 4$ in. Bevel gears E, F, and G have the same number of teeth.

Determine: Velocity of slider S.

Solution. This mechanism is used to convert rotary motion into reciprocating motion. Compared to the conventional slider crank mechanism, this design reduces shock (owing to a reduction in acceleration) at each end of the stroke.

Hollow shaft H is integrally formed to gear E and crank slider M. Concentric to the hollow shaft is a central shaft S. One end of S is fastened to a short crank AB, and the other end is pinned to gear F. Links BC and CD are connected to a pivot pin at C. Bevel gear G drives gears E and F at the same rpm in opposite directions. Thus, AB and M rotate at the same rpm in opposite directions.

STEP 1. Refer to Fig. 4.25(b). AB rotates ccw and M rotates cw at 1000 rpm.

STEP 2. Label contact points at C, X on L and Y on M.

STEP 3. Calculate V_B:

$$V_B = AB\omega_{AB} = .5[1000\,(2\pi)] = 3142 \text{ in./min}$$

STEP 4. Let $K_V = 3142$ in./min/.5 in. Lay out V_B.

STEP 5. Construct V_Y by the proportion

$$\frac{|V_Y|}{|V_B|} = \frac{AY}{AB}$$

where V_Y is constructed perpendicular to AY at Y.

STEP 6. Construct the velocity component of V_X along XB (or CB) where $_{CB}V_B = {}_{XB}V_X$.

STEP 7. Draw a line perpendicular to XB which passes through the tip of vector $_{XB}V_X$.

STEP 8. Determine V_X. Construct a TT line which is parallel to the slot in M and passes through the tip of vector V_Y. The intersection of the TT line and the perpendicular line (step 7) locates V_X. V_X extends from X to the intersection of the lines.

STEP 9. Determine V_D by components:

$$V_D = 4340 \text{ in./min}$$

Hence the slider velocity is 4340 in./min.
Note that in this problem V_X was determined by two sets of components:

$$V_X = {}_{XB}V_X +\!\!\!\!\!\to {}_{\perp XB}V_X$$

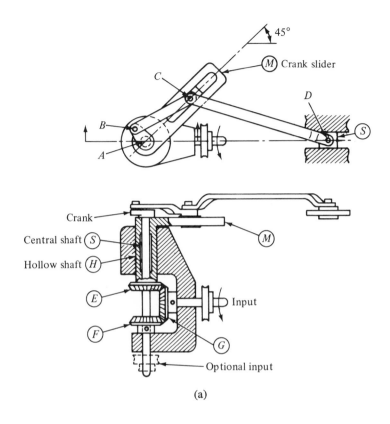

(a)

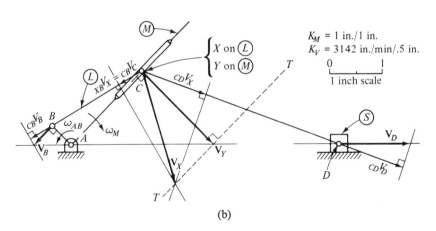

(b)

Fig. 4.25. (a) *Rotating slider crank mechanism.* (b) *Simplified kinematic skeleton.* (Courtesy of B. H. Wallace of Wallace Automation, Inc., U.S. Patent 3,134,266.)

125

and

$$V_X = {}_{NN}V_X \mathbin{+\!\!\!\!\!+} {}_{TT}V_X$$

where ${}_{NN}V_X = V_Y$. To avoid cluttering Fig. 4.25(b), the velocities ${}_{\perp XB}V_X$, ${}_{NN}V_X$, and ${}_{TT}V_X$ were not shown.

4.14 Relative Velocity Between Two Points on the Same Body

The only method shown so far for determining the angular velocity of a body is the instant (or permanent center) method. Another method quite frequently used, especially when the instant center is located far off the mechanism, is the relative velocity method. This method utilizes the relative linear velocity of two points on the same body.

The relative velocity method can best be explained by an example. Figure 4.26(a) shows a wheel $\textcircled{W}$ rolling without slipping on surface $\textcircled{S}$. V_A and V_B are known. The angular velocity of $\textcircled{W}$ will be determined from

$$\omega_W = \frac{V_{B/A}}{AB} \tag{4.10}$$

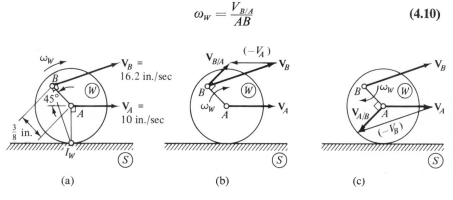

(a)　　　　　　　　　　　　(b)　　　　　　　　　　　　(c)

Fig. 4.26. (a) V_A and V_B are known. (b) $V_{B/A} = V_B \mathbin{+\!\!\!\!\!+} (-V_A)$. (c) $V_{A/B} = V_A \mathbin{+\!\!\!\!\!+} (-V_B)$.

where the symbol $V_{B/A}$ means the velocity of B relative to A. This can be physically interpreted as the velocity of B as seen by an observer on point A. Further, the observer on A must have a reference body in order to judge the motion of B. In this example $\textcircled{S}$ is the reference body. The construction of $V_{B/A}$ is shown in Fig. 4.26(b):

$$V_{B/A} = V_B \rightarrow V_A$$

or

$$V_{B/A} = V_B \mathbin{+\!\!\!\!\!+} (-V_A) = 7.5 \text{ in./sec}$$

Note that $V_{B/A}$ is perpendicular to AB at B. Now, substitute $V_{B/A}$ into Eq. (4.10):

$$\omega_W = \frac{7.5 \text{ in./sec}}{\frac{3}{8} \text{ in}} = 20 \frac{\text{rad}}{\text{sec}} \text{ cw}$$

There is cw rotation because it would appear to the observer on A that line AB rotates cw.

If desired, $V_{A/B}$ can be used to determine ω_W; refer to Fig. 4.26(c):

$$\omega_W = \frac{V_{A/B}}{AB} = \frac{7.5 \text{ in./sec}}{\frac{3}{8} \text{ in.}} = 20 \frac{\text{rad}}{\text{sec}} \text{ cw}$$

PROBLEMS

4.1 Figure P4.1 shows two photographs taken with a high-speed camera at 1200 frames/sec. The fifth and twentieth frames are shown. Determine the average angular velocity for the lever. The lever is shown in the center of each photograph.

Fig. P4.1. (Courtesy of Pitney-Bowes, Inc.)

4.2 Figure P4.2(a) shows a four-bar linkage $ABCD$ used in an electric shaver. AB is the driver and CD is the follower. Figure 4.2(b) is a multiple-exposure photograph obtained with a Strobotac set at 18,000 flashes/min. Determine the average angular velocity of CD between flashes. Use the edge of the follower counterweight to determine angular displacement.

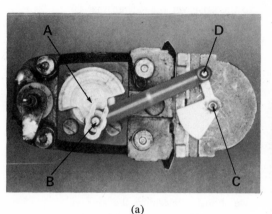

(a)

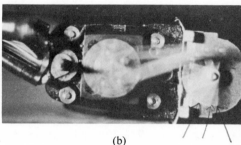

(b)

Fig. P4.2. (Courtesy of Sunbeam Corporation.)

4.3 A four-station Geneva mechanism, Fig. P4.3, is used to produce intermittent rotary motion. Disk D rotates at a constant 60 rpm cw. A pin P, which is rigidly fastened to D, drives disk F ccw. $AP = 2$ in., and $AB = 2.83$ in. Approximate the angular velocity of F when AP is 15 deg above the horizontal. Use $\omega \approx \omega_{av} = \Delta\theta/\Delta t$.

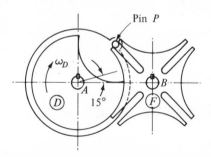

Fig. P4.3.

4.4 Figure P4.4 shows a gear slider mechanism. This mechanism can produce a variety of cyclic output motions. The input is into crank AB, and the output is from gear F. Crank $AB = 2$ in., the connecting rod $BCD = 6$ in., the gear diameter $F = 1.5$ in., and the gear diameter $G = 2.5$ in. The connecting rod is pinned to gear G at B and C. The angular velocity of AB is 100 rad/sec ccw. Approximate the angular velocity of F at this instant. Use $\omega \approx \omega_{av} = \Delta\theta/\Delta t$.

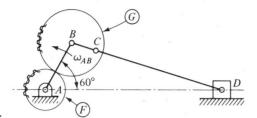

Fig. P4.4.

4.5 Determine the approximate linear velocity for point P in Prob. 4.3 when AP is 15 deg above the horizontal. Use $V \approx V_{av} = \Delta S/\Delta t$.

4.6 Determine the approximate linear velocity for point D in Prob. 4.4. Use $V \approx V_{av} = \Delta S/\Delta t$.

4.7 Line OX is on a wheel which rotates ccw about center O. The angular displacement of OX from a horizontal reference line is given as $\theta = 2t^2 + 3t + 1$, where θ is in radians and t is in seconds. (a) Plot θ vs. t (θ on ordinate axis). (b) Approximate the angular velocity at 5 sec. *Hint:* Determine $\Delta\theta/\Delta t$ between 4.99 and 5.01 sec. (c) Calculus shows that the angular velocity equation is $\omega = 4t + 3$, where ω is in radians per second and t is in seconds. Determine ω at 5 sec, compare your answer with part b.

4.8 In Fig. P4.8, a link $\textcircled{L}$ rotates 600 rpm cw about point O. Determine the linear velocity of points P and Q.

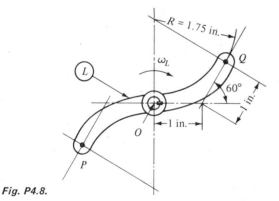

Fig. P4.8.

4.9 An eccentric cam rotates at 300 rpm cw (Fig. P4.9). The eccentricity $OC = .5$ in., and the cam diameter $= 2$ in. (a) Determine the linear velocity for points P and Q at this instant. (b) Determine the velocity of a point, on the cam surface, which has the largest velocity.

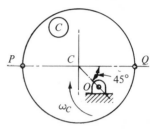

Fig. P4.9.

4.10 It is desired to mill a piece of machine steel at a cutting speed of 80 ft/min. The cutter diameter is 3 in. (a) Determine the cutter rpm using $V = R\omega$. (b) An approximate equation often used to calculate rpm is $n = 4(C.S.)/D$, where $n = $ rpm, C.S. $= $ the cutting speed (ft/min), and $D = $ diameter (in.). Calculate rpm using this equation. Compare answers and determine the reason for the difference.

4.11 Figure P4.11 shows a floating link AB, where $V_A = 10$ ft/sec and $V_B = 17.32$ ft/sec. Are these velocities reasonable? Why?

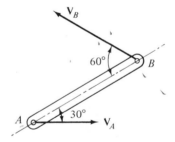

Fig. P4.11.

4.12 Figure P4.12 shows the cross section of a single-cylinder, two-cycle gasoline engine. Intake and exhaust ports are located in the cylinder wall near bottom dead center. Determine the piston velocity when the crank is 30 deg from BDC. The crank $= .875$ in., the connecting rod $= 3.125$ in., and the engine speed is 3000 rpm.

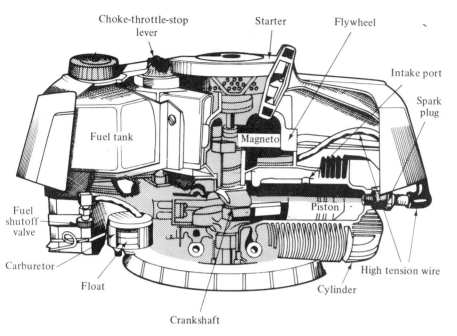

Fig. P4.12. (Courtesy of Jacobsen Manufacturing Co.)

4.13 Figure P4.13 shows the mechanism of a portable carving knife. The gear G rotates about O. Molded onto G is a pin which is located $\frac{3}{16}$ in. from O. This pin is part of a rotating slider. Determine the maximum knife velocity when G rotates at 5 rev/sec ccw.

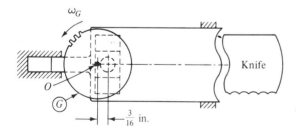

Fig. P4.13. (Courtesy of Roto Broil Corp.)

4.14 Figure P4.14 shows a photograph and the kinematic skeleton of a machine used to determine whether or not color may be transferred (by rubbing) from the surface of a dyed textile material to a piece of white crock cloth. The testing procedure and evaluation is covered in American Standards Association specification ASA L 14.72-1963. For the mechanism position shown, determine the rubbing velocity between the tested material and the crock cloth. The hand crank is rotated at 1 rev/sec ccw. The dimensions are $AB = 4.4$ in., $BC = 2.2$ in., $CD = 10.9$ in., $ED = .4$ in., and $EF = 16.1$ in.

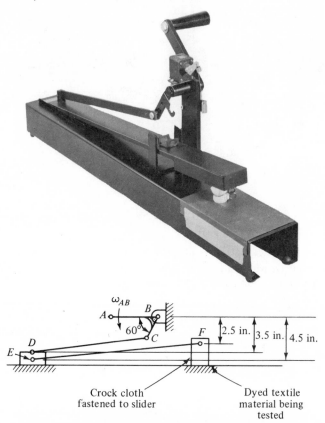

Fig. P4.14. *Hand-operated Crockmeter simulates the rubbing motion of a human finger and forearm.* (Courtesy of Atlas Electric Devices Co.)

4.15 Figure P4.15 shows a 150-ton scrap-charging machine. The tilt mechanism consists of two parallel-operated four-bar link mechanisms (only one of the four-bar mechanisms can be clearly seen in the photograph). Determine the angular velocity of CD when crank AB is horizontal. AB rotates at 2 rpm cw. $AB = 3$ ft, $1\frac{1}{2}$ in.; $BC = 8$ ft, $6\frac{1}{2}$ in.; $AC = 7$ ft, 3 in.; $CD = 8$ ft, 6 in.; and $AD = 9$ ft, 6 in.

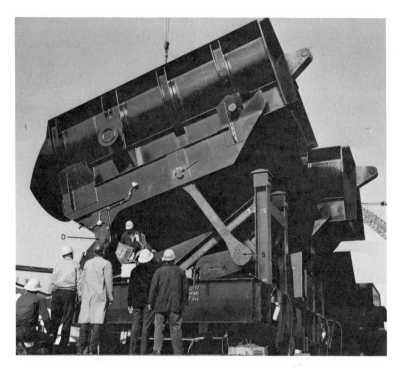

Fig. P4.15. Tilt mechanisms are powered by a 100-hp electric motor through a multistage gear-reduction unit (267:1) providing up to $5\frac{1}{2}$ million in.-lb of torque. (Courtesy of Dravo Corp.)

4.16 Figure P4.16 shows a simplified kinematic drawing of a 400-ton knuckle joint press when the ram is at bottom position. This press uses two toggle linkages to obtain a high mechanical advantage. If AB rotates at 265 rpm ccw, determine the ram velocity at midstroke (ram moving downward). $AB = 2.7$ in., $BC = 22.6$ in., and $CD = CE = 11.5$ in.

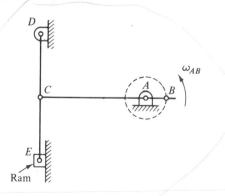

Fig. P4.16. (Courtesy of E. W. Bliss Co.)

4.17 In Fig. P4.17, crank AB rotates at 100 rpm cw. Link DCE connects gear centers. Determine the velocity of the moving rack. $AB = 2$ in., $BC = 6$ in., and the gear diameters $= 1$ in.

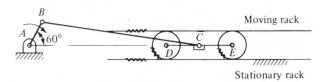

Fig. P4.17.

4.18 In Fig. P4.18, $V_B = 20$ ft/min. Determine V_F and ω_{EF} for the position shown. The dimensions are $AB = .5$ in., $BEC = 2.25$ in., $CD = 1.0$ in., $DA = 2.0$ in., $BE = 1.0$ in., $EF = 1.0$ in., $FG = 2.0$ in., and $DG = 2.5$ in.

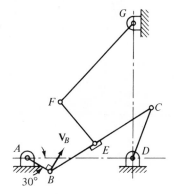

Fig. P4.18.

4.19 Scale Fig. P4.19 from text. Assume a reasonable length for V_B. Determine (a) the linear velocity for points C and D and (b) the angular velocity for BC, CD, and the gear. Angular velocities are to be in symbolic form; for example, $\omega_{BC} = V_B/BI_{BC}$ cw.

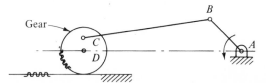

Fig. P4.19.

4.20 Scale Fig. P4.20 from text. Assume a reasonable length for V_B. Determine (a) the linear velocity for points C, E, and F and (b) the angular velocity for BC, BE, and CD. Angular velocities are to be in symbolic form; for example, $\omega_{BC} = V_C/CI_{BC}$.

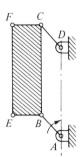

Fig. P4.20.

4.21 Scale Fig. P4.21 from text. Assume a reasonable length for V_B. Determine (a) the linear velocity for points C, D, and E and (b) the angular velocity for BC, CD, DE, EF, AF, and the slider. Angular velocities are to be in symbolic form; for example, $\omega_{BC} = V_B/BI_{BC}$ cw.

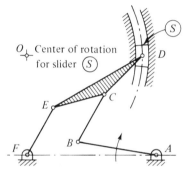

Fig. P4.21.

4.22 Scale Fig. P4.22 from text. Assume a reasonable length for V_B. Determine (a) the linear velocity for points C and E and (b) the angular velocity for BC, EB, and CD. Angular velocities are to be in symbolic form; for example, $\omega_{BC} = V_E/EI_{BC}$ cw.

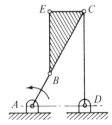

Fig. P4.22.

4.23 In Fig. P4.23, determine the velocity of the slider for each 30 deg of rotation for crank AB; start at the 0-deg mark. The crank rotates ccw at a constant angular velocity. $AB = .5$ in., and $BC = 3.0$ in. It is required that the magnitude for each V_C be laid out perpendicular to the tail of vector V_C, as indicated in the figure. Plot magnitudes above the slider when the slider is moving to the left and below when the slider is moving to the right. When all the magnitudes have been established, including extreme positions, draw a smooth curve through the end points; the resulting figure is commonly called a velocity diagram. Answer the following questions: (a) Where is the crank when the slider reaches its maximum velocity? (b) Is the time for the forward and reverse strokes the same? Why? (c) If the linear velocity for B is 2000 ft/min, what is the velocity for the slider at the 45-deg mark? Use the velocity diagram.

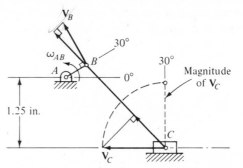

Fig. P4.23.

4.24 Determine the tip velocity (for points P and Q) of a helicopter blade relative to the earth. Each blade length is 30 ft. The rotor rotates at 234 rpm ccw, and the helicopter flight speed is 100 mph relative to the earth. Refer to Fig. P4.24.

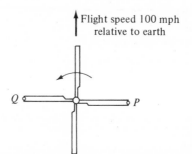

Fig. P4.24. (Courtesy of Sikorsky Aircraft, Division of United Aircraft Corp.)

4.25 A nonslip belt connects pulleys Ⓐ, Ⓑ, and Ⓒ; refer to Fig. P4.25. Pulley Ⓐ rotates 1 rad/min ccw. Spur gear Ⓓ drives Ⓔ. Gear Ⓓ is fixed to Ⓒ. The pitch diameters are Ⓐ = 1.50 in., Ⓑ = .75 in., Ⓒ = .75 in., Ⓓ = .50 in., and Ⓔ = 1.00 in. Determine (a) the belt velocity (ft/sec), (b) the angular velocity for pulley Ⓑ, and (c) the angular velocity for gear Ⓔ.

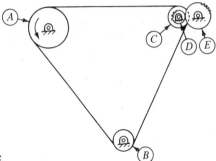

Fig. P4.25.

4.26 In Fig. P4.26, cam C moves horizontally to the right, pushing up roller R and slider S. R and S are pin-connected at point P. If C moves at 1 in./sec, determine the velocity of S. The roller diameter is 1 in. Assume pure rolling.

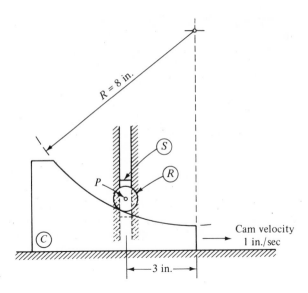

Fig. P4.26.

4.27 Figure P4.27 shows a "squeeze-type" pump. As the arm rotates, a pair of rotating rollers squeeze fluid through a flexible tube. Arm AOB rotates 500 rpm. The roller diameters are 1 in. and $OA = OB = 1\frac{1}{2}$ in. Determine (a) the maximum fluid velocity and (b) the rpm of the roller. Assume pure rolling.

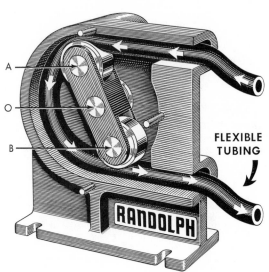

FLEXIBLE TUBING

Fig. P4.27. (Courtesy of the Randolph Co.)

4.28 Figure P4.28 shows a rateometer, a machine used to increase reading speed. A synchronous electric motor drives a cone at a constant angular speed of 1 rev/15 sec. The cone moves a T-shank over a serrated base plate. Traction between the cone and T-shank is increased by a hollow rubber tube which is fastened to the T-shank. The cone is free to rotate at point O. $OA = OB = 7\frac{1}{2}$ in., and $AB = 3\frac{3}{8}$ in. (a) What is the reading rate (words/min) when the T-shank is positioned 2 in. from O? The reading material has a word count of 40 words/in. Assume no slip. (b) Determine a general equation for the reading rate (words/min) at any position Z of the T-shank. Let $Z =$ distance in inches and $W =$ words per inch.

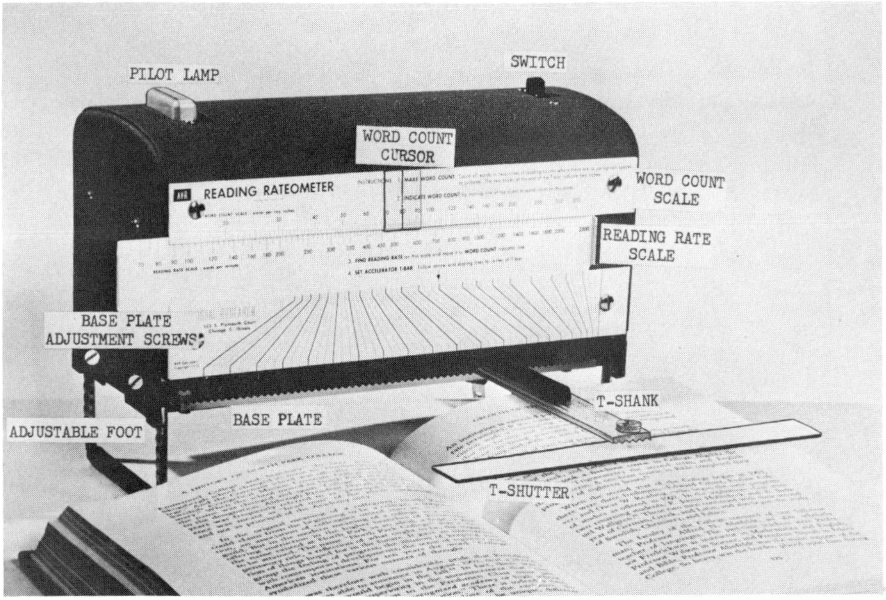
PILOT LAMP
SWITCH
WORD COUNT CURSOR
READING RATEOMETER
WORD COUNT SCALE
READING RATE SCALE
BASE PLATE ADJUSTMENT SCREWS
T-SHANK
BASE PLATE
ADJUSTABLE FOOT
T-SHUTTER

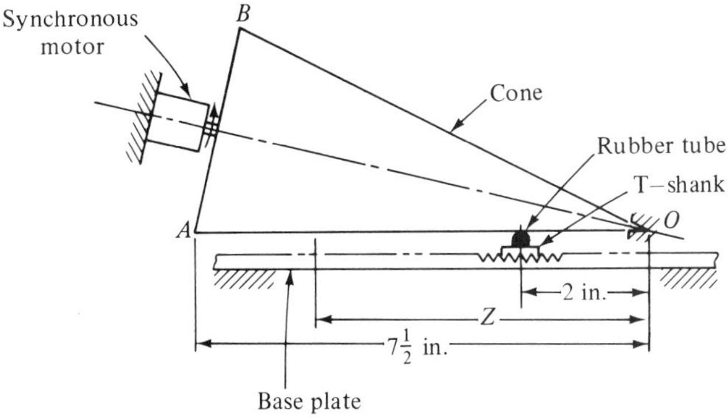
Synchronous motor
B
Cone
Rubber tube
T–shank
A
O
2 in.
Z
$7\frac{1}{2}$ in.
Base plate

Fig. P4.28. (Courtesy of Audio-Visual Research.)

4.29 Figure P4.29 shows a simplified drawing of a planetary gear train. The input velocity into the arm is 3.64 rad/sec ccw (viewed from the right side). Gear Ⓐ is stationary. The arm moves gears Ⓑ and Ⓒ around gears Ⓐ and Ⓓ. Gears Ⓑ and Ⓒ are keyed onto shaft Ⓢ. The gear diameters are Ⓐ = 2.5 in., Ⓑ = 3.0 in., Ⓒ = 2.5 in., and Ⓓ = 3.0 in. Determine (a) the absolute angular velocity for gear Ⓑ and (b) the magnitude and direction of rotation for gear Ⓓ.

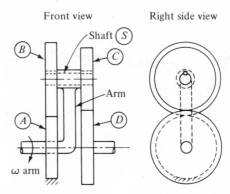

Fig. P4.29.

4.30 In Prob. 4.29, change the diameter of gear Ⓒ to 3.5 in. and the diameter of gear Ⓓ to 2 in. Determine answers to parts a and b as spelled out in Prob. 4.29.

4.31 to 4.34 Scale Figs. P4.31 to P4.34 from text. Assuming a reasonable velocity scale, determine the velocity of point F. Note that ω_L is cw.

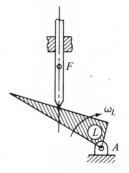

Fig. P4.31.

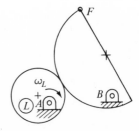

Fig. P4.32.

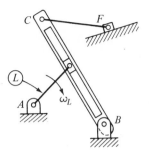

Fig. P4.33.

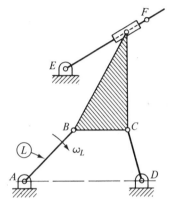

Fig. P4.34.

4.35 A car is attempting to ascend an icy hill. The rear wheels are rotating at 14 rev/sec; however, the car is moving downhill at 7 ft/sec. Determine the sliding velocity between the tire and the ice. The tire diameter is 2 ft.

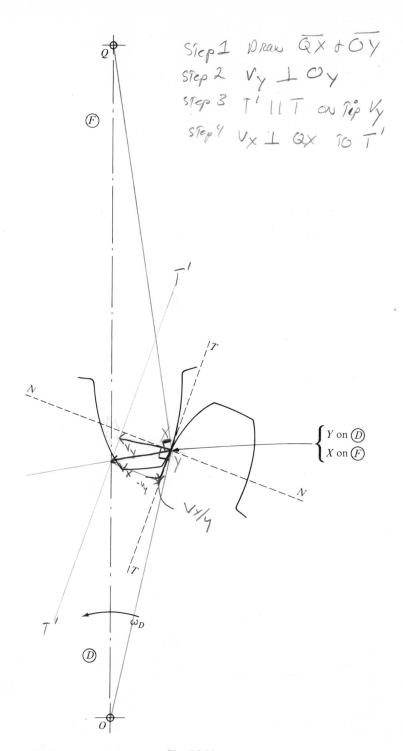

Step 1 DRAW $\overline{QX}$ & $\overline{OY}$
Step 2 $V_y \perp OY$
Step 3 $T' \parallel T$ on Tip V_y
Step 4 $V_x \perp QX$ TO T'

$\begin{cases} Y \text{ on } \textcircled{D} \\ X \text{ on } \textcircled{F} \end{cases}$

Fig. P4.36.

142

4.36 Figure P4.36 shows a pair of gear teeth in contact. Contact takes place along the pressure line or normal line *NN*. Driver Ⓓ rotates at 1000 rpm ccw about *O*, and gear Ⓕ rotates about *Q*. Scale dimensions directly from text, let $K_M = 1$ in./1 in. Determine (a) the linear velocity of point *Y*, (b) the angular velocity of Ⓕ, and (c) the sliding velocity ($V_{X/Y}$).

4.37 Figure P4.37 shows a sketch and kinematic drawing of a vane-type air motor. Sealing is accomplished by centrifugal force holding the vanes tightly against the housing. Determine the sliding velocity between the vane and rotor at the 30-deg position shown in the figure. The rotor rotates at 8000 rpm cw.

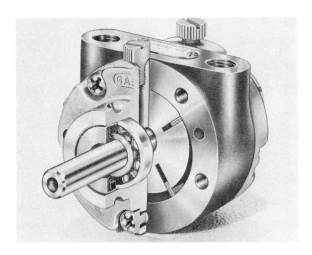

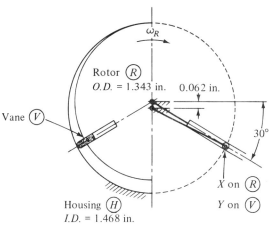

Fig. P4.37. (Courtesy of Gast Manufacturing Corp.)

4.38 Figure P4.38, determine the linear velocity of point P at the 45-deg position. The piston moves at a constant 5 in./sec. *Hint:* Use a relative velocity equation where 5 in./sec is the piston velocity relative to the cylinder.

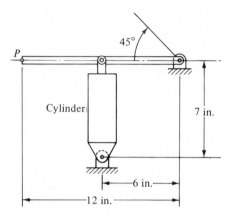

Fig. P4.38.

4.39 Determine the exact angular velocity for gear Ⓕ in Prob. 4.4.

4.40 At this instant gear Ⓖ is rotating at 2 rpm ccw, Fig. P4.40. Points B and D are fastened to the gears as indicated. Determine the velocity of point C, using two components. The pitch diameter for gear Ⓖ $= 1.5$ in., the pitch diameter for Ⓗ $= 2.5$ in., $BC = 1.5$ in., and $CD = 2.5$ in.

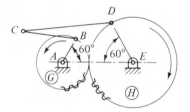

Fig. P4.40.

4.41 Scale Fig. P4.41 from text. Assume a reasonable length for V_B. Determine the linear velocity for points C, F, E, and D. Use two components to determine the velocity of point D.

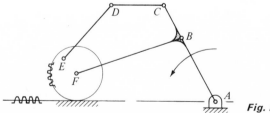

Fig. P4.41.

Chapter **5**

ACCELERATION

5.1 Introduction

When a mechanism is operating at high speed, the dynamic forces impressed upon the various members of the mechanism may be quite high. It was Newton who discovered that the dynamic force is related to the linear acceleration by the simple equation $F = mA$, where F = force, m = mass, and A = acceleration. Thus a dynamic force analysis of a mechanism first requires an acceleration analysis. It is the intent of this chapter to introduce some of the methods used in acceleration analysis.

5.2 Angular Acceleration

Angular acceleration, like angular displacement and angular velocity, relates to the motion of a line. Angular acceleration is defined as the time rate of change of angular velocity; that is,

$$\text{Angular acceleration} = \frac{\text{change in angular velocity}}{\text{change in time}}$$

145

The symbol α (alpha) will be used to identify angular acceleration. Average angular acceleration is

$$\alpha_{av} = \frac{\Delta\omega}{\Delta t} = \frac{\omega_2 - \omega_1}{t_2 - t_1}$$

and the instantaneous angular acceleration is

$$\alpha = \lim_{\Delta t \to 0} \frac{\Delta\omega}{\Delta t} = \frac{d\omega}{dt} \tag{5.1}$$

For plane motion, α can be treated as a scalar quantity, being either cw or ccw. Units for angular acceleration are rpm/sec, rad/sec/min, rad/sec/sec, or simply rad/sec^2.

Figure 5.1 illustrates the physical significance of angular acceleration. Here

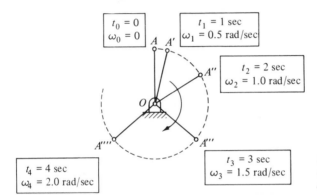

Fig. 5.1. Link OA rotates at a constant angular acceleration of .5 rad/sec^2 cw.

several instantaneous positions of OA are shown. Link OA is rotating cw about O. Observe that for each successive second the angular velocity is increasing by .5 rad/sec, and also that the angular displacement between adjacent positions is increasing. Link OA is said to be accelerating or speeding up at .5 rad/sec/sec or .5 rad/sec^2 cw. We are assuming here that the angular acceleration is constant.

When ω and α act in the same direction, as in Fig. 5.1, the link or body is accelerating or speeding up. If ω and α are in opposite directions, a deceleration or slowing down is indicated.

EXAMPLE 5.1

Given: A large wheel starts from rest and is accelerated at 1000 rpm/min about its center.

Determine: Angular velocity and angular distance moved (in revolutions) after 2.5 min.

Solution. The rate of change of angular velocity is 1000 rpm for each minute; thus, 1 min from rest the wheel has an angular velocity of 1000 rpm. Two minutes from rest it would be 2000 rpm, etc. Mathematically,

$$\omega = \alpha t \tag{1}$$

For 2.5 min,

$$\omega = \left(1000 \, \frac{\text{rpm}}{\text{min}}\right)(2.5 \text{ min}) = 2500 \text{ rpm}$$

Equation (1) is a linear equation where α is the slope of the line; refer to Fig. 5.2.

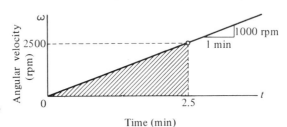

Fig. 5.2. *Angular velocity-time diagram.*

     The angular distance moved in 2.5 min is the area under the curve bounded by $t = 0$ and $t = 2.5$ min. This is the area of a triangle with a height of 2500 rev/min and a base of 2.5 min. The angular distance is

$$\theta = \frac{1}{2}\left(2500 \, \frac{\text{rev}}{\text{min}}\right)(2.5 \text{ min}) = 3125 \text{ rev}$$

For this particular type of problem the angular distance can be generally expressed as

$$\theta = \tfrac{1}{2}\omega t$$

or, substituting for ω; use Eq. (1),

$$\theta = \tfrac{1}{2}\alpha t^2$$

It is interesting to note that at the instant of starting $t = 0$, $\omega = 0$, and $\theta = 0$ but $\alpha = 1000$ rpm/min.

5.3 Linear Acceleration

     The linear acceleration of a point or particle is defined as the change in linear velocity with respect to time. Where the average linear acceleration is

$$A_{\text{av}} = \frac{\Delta V}{\Delta t}$$

the instantaneous acceleration is

$$A = \lim_{\Delta t \to 0} \frac{\Delta V}{\Delta t} = \frac{dV}{dt}$$

Acceleration is a vector quantity which has the same sense as vector ΔV, or dV in the case of instantaneous acceleration. Units for linear acceleration are in./sec², ft/sec², etc.

If a point moves with increasing or decreasing velocity in rectilinear translation [Fig. 5.3(a)], the acceleration vector will be along the line of motion. That is, the velocity and acceleration are along the same line.

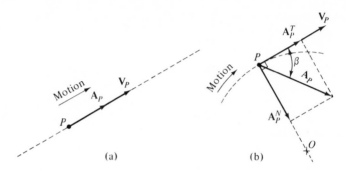

Fig. 5.3. *Instantaneous acceleration A_P for a point P moving with increasing velocity in (a) rectilinear translation and (b) curvilinear motion.*

The instantaneous acceleration for a point which moves with other than rectilinear translation is usually at some angle to the instantaneous velocity. To have an acceleration, the point need not be speeding up or slowing down. Generally the acceleration of a point is determined by rectangular components which are normal and tangent to the path of motion. In Fig. 5.3(b) the normal acceleration (relative to the earth) is A_P^N. This component is always directed toward the center of curvature O. The tangential acceleration (relative to the earth) is A_P^T; it is directed in the same direction as V_P when the point is speeding up, and opposite to V_P when the point is slowing down. The acceleration of P (relative to the earth) is the vector sum of the components; namely, $A_P = A_P^N + A_P^T$.

EXAMPLE 5.2

Given: An automobile starts from rest and attains 60 mph in 6 sec.

Determine: Acceleration of the automobile in ft/sec².

Solution. Assuming the acceleration to be constant, the velocity equation is

$$V = At$$

Thus

$$A = \frac{V}{t} = \frac{60 \text{ mph}}{6 \text{ sec}} \left(\frac{1 \text{ hr}}{3600 \text{ sec}}\right) \left(\frac{5280 \text{ ft}}{\text{mile}}\right) = 14.7 \frac{\text{ft}}{\text{sec}^2}$$

5.4 Normal Acceleration

In Fig. 5.4 (a), link OP rotates cw about O at a constant angular velocity, and the angular acceleration is zero. There are two positions shown for the link, that at OP and OP_1. The linear velocity at positions P and P_1 are equal in magnitude but differ in direction. To establish the linear acceleration for point P at position P, we must first determine the velocity change or difference between V_P and V_{P_1}. Thus

$$\Delta V = V_{P_1} \rightarrow V_P$$

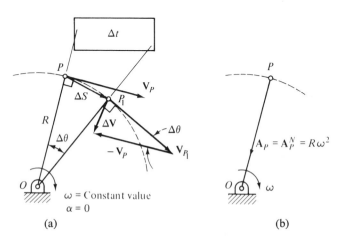

Fig. 5.4. (a) Link OP rotates cw at a constant angular velocity. Two positions of the link are shown: at OP and OP₁. (b) The instantaneous linear acceleration for point P, at position OP, is $A_P = A_P^N$ and is directed toward point O.

This vector subtraction is shown in Fig. 5.4(a) at position P_1. Since $-V_P$ is perpendicular to OP, the angle between $-V_P$ and V_{P_1} is equal to the angular displacement from OP to OP_1, namely $\Delta\theta$. The vector triangle at P_1 is an isosceles triangle which is similar to triangle POP_1. Thus

$$\frac{\Delta V}{V} = \frac{\Delta S}{R} \quad \text{and} \quad \Delta V = \frac{V(\Delta S)}{R}$$

where $V = |V_P| = |V_{P_1}|$, $\Delta S = |\overline{PP_1}|$, and $R = OP = OP_1$. The average linear acceleration of point P when moving from P to P_1 is

$$A_{P_{av}} = \frac{\Delta V}{\Delta t} = \frac{V}{R}\left(\frac{\Delta S}{\Delta t}\right) \tag{1}$$

To obtain the instantaneous acceleration at position P, let Δt approach zero in Eq. (1); that is, consider ΔS and Δt as infinitesimals dS and dt. Recall that

$dS/dt = V$; then Eq. (1) becomes

$$A_P = \frac{V^2}{R} \tag{2}$$

Acceleration is a vector quantity which points in the same direction as dV. As Δt and $\Delta\theta$ approach zero, the direction for ΔV approaches center O. Finally, in the limit, dV points at O. Thus the instantaneous acceleration A_P (at position P) can be represented graphically as a vector which points toward O and is along radial line OP; see Fig. 5.4(b). This acceleration is labeled the radial or normal acceleration and is generally symbolized as A^N. In this discussion, $A_P = A_P^N$.

There are two other expressions for the normal acceleration:

$$A^N = \frac{V^2}{R} = \frac{(R\omega)^2}{R} = R\omega^2 \quad \text{and} \quad A^N = \frac{V^2}{R} = V\left(\frac{V}{R}\right) = V\omega$$

In summary, the normal acceleration is calculated by

$$A^N = R\omega^2 = \frac{V^2}{R} = V\omega \tag{5.2}$$

Typical units for Eq. (5.2) are

$$A^N = \text{normal acceleration, ft/sec}^2$$
$$R = \text{radius, ft}$$
$$\omega = \text{angular velocity, rad/sec}$$
$$V = \text{velocity, ft/sec}$$

Normal acceleration is due to the change in velocity direction, and is directed toward point O. It is important to recognize that the acceleration A_P is relative to the earth or to point O which is on earth; thus $A_P = A_{P/O}$.

EXAMPLE 5.3

Given: A link (Fig. 5.5) rotates at a constant 1000 rpm cw about O. $OA = 3$ in., and $OB = 1.5$ in.

Determine: Acceleration for points A and B at the position shown.

Solution. Since the angular velocity is constant, the acceleration will be all normal. Using Eq. (5.2),

$$A_A = A_A^N = \overline{OA}\omega^2 = 3[(1000)2\pi]^2 = 118.2 \times 10^6 \text{ in./min}^2$$

The normal acceleration for point B is half of A_A because $\overline{OB} = \frac{1}{2}(\overline{OA})$:

$$A_B = A_B^N = 59.1 \times 10^6 \text{ in./min}^2$$

$K_M = 1$ in./1 in.

$K_A = 118.2 \times 10^6$ in./min^2/1 in.

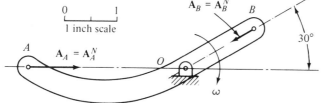

Fig. 5.5. *Instantaneous linear acceleration for points* A *and* B *on a link rotating at constant rpm.*

Accelerations are laid out to a scale of $K_A = 118.2 \times 10^6$ in./min^2/1 in. and are directed toward point O.

5.5 Normal and Tangential Acceleration

The general case for the linear acceleration of a point on a rotating link occurs when the link has an angular acceleration.

Figure 5.6(a) shows a link OP rotating cw about O with an angular acceleration in the same direction as the angular velocity. Two positions are shown for the link: at OP and OP_1. Since the angular velocity is increasing, the linear velocity at P_1 is larger than that at P. The change in linear velocity is shown at position P_1, where $\Delta V = V_{P_1} \rightarrow V_P$. Vector ΔV is resolved into components ΔV^T and ΔV^N, where ΔV^T is along V_{P_1} and is the algebraic difference between V_{P_1} and V_P. This quantity represents the change in velocity magnitude. Vector

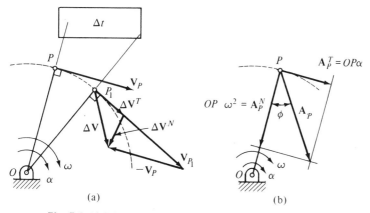

Fig. 5.6. *Link* OP *rotates cw with an angular acceleration. The instantaneous linear acceleration for point* P, *at position* OP, *[Fig. 5.6. (b)] is* $A_P = A_P^T + A_P^N$.

ΔV^N is due to the change in velocity direction. Since ΔV^N corresponds to ΔV in Fig. 5.4(a), there is no need to discuss this quantity again; therefore, we shall concentrate our attention on component ΔV^T.

The algebraic expression for ΔV^T is

$$\Delta V^T = V_{P_1} - V_P = R\omega_1 - R\omega = R(\omega_1 - \omega) \tag{1}$$

where $R = OP = OP_1$, ω_1 represents the angular velocity at position OP_1, ω represents the angular velocity at position OP. Equation (1) can be simply expressed as

$$\Delta V^T = R\,\Delta\omega$$

$\Delta\omega$ is the change in angular velocity. The average linear acceleration (related to component ΔV^T) of point P when moving from P to P_1 is

$$A^T_{P_{av}} = \frac{\Delta V^T}{\Delta t} = \frac{R\,\Delta\omega}{\Delta t} \tag{2}$$

The instantaneous acceleration at position P is obtained by letting Δt approach zero; thus Eq. (2) becomes

$$A^T_P = \frac{R\,d\omega}{dt} = R\alpha$$

where angular acceleration α is the instantaneous value at position OP. This component of the linear acceleration is called the tangential acceleration. It is directed tangentially to the path of motion or perpendicularly to radial line OP, and its direction is consistent with that of α.

The tangential acceleration is generally expressed as

$$A^T = R\alpha \tag{5.3}$$

Typical units are

$$A^T = \text{tangential acceleration, ft/sec}^2$$
$$R = \text{radius, ft}$$
$$\alpha = \text{angular acceleration, rad/sec}^2$$

In conclusion, the instantaneous linear acceleration for point P at position OP consists of two components, A^T_P and A^N_P [see Fig. 5.6(b)], where A^T_P represents the acceleration due to the change in magnitude of velocity, and A^N_P results from a change in inclination of velocity. The resultant is

$$A_P = A^T_P + A^N_P$$

Since the components are at right angles to each other, A_P can be determined algebraically as

$$A_P = \sqrt{(A_P^T)^2 + (A_P^N)^2}$$

and angle ϕ, between A_P and OP is

$$\tan \phi = \frac{A_P^T}{A_P^N} = \frac{\overline{OP}\alpha}{\overline{OP}\omega^2} = \frac{\alpha}{\omega^2}$$

5.6 Gear Accelerations

It was stated in Sect. 4.12 that the peripheral speed (on the pitch circle) for two contacting spur gears are equal. This is true whether the gears are speeding up, slowing down, or rotating at a constant angular velocity. Because of this speed synchronization between gears, all the points on the pitch circle, of either gear, have the same "peripheral" acceleration. That is, in Fig. 5.7 the magnitude of the tangential acceleration for points X, Y, P, and Q are equal. Note that $A_X \neq A_Y \neq A_P \neq A_Q$. At the point of contact between gears,

$$A_Y^T = A_X^T$$

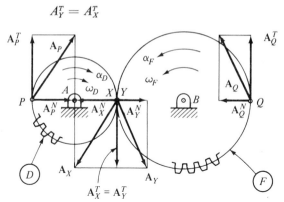

Fig. 5.7. *Acceleration and component accelerations for* X, Y, P, *and* Q.

Using Eq. (5.3), this equality becomes

$$R_F\alpha_F = R_D\alpha_D$$

Rearranging,

$$\frac{\alpha_F}{\alpha_D} = \frac{R_D}{R_F} \tag{1}$$

Combine Eq. (1) with the angular velocity relationship,

$$\frac{\omega_F}{\omega_D} = \frac{R_D}{R_F}$$

we have

$$\frac{\alpha_F}{\alpha_D} = \frac{\omega_F}{\omega_D} = \frac{R_D}{R_F} \tag{5.4}$$

Equation (5.4) shows that the instantaneous angular accelerations and angular velocities are inversely proportional to their radii.

EXAMPLE 5.4

Given: At this instant, gear ⒟ of a three-gear drive (Fig. 5.8) rotates at 300 rpm ccw and is accelerating at 90 rad/sec² ccw. Wheel Ⓦ and gear Ⓕ are pinned together. The diameters are ⒟ = 1 in., Ⓔ = 1.5 in., Ⓕ = 3 in., and Ⓦ = 4 in.

Determine: The linear acceleration for point Q on the circumference of wheel Ⓦ.

Solution. In Fig. 5.8, label contact points X, Y, Z, and P. At this instant, $V_X = V_Y$ and $V_Z = V_P$; however, since $|V_Y| = |V_Z|$, then $|V_X| = |V_P|$. Thus the magnitude relationship for velocity is

$$V_X = V_P$$

or

$$R_D \omega_D = R_F \omega_F$$

Solving for ω_F,

$$\omega_F = \omega_D \frac{R_D}{R_F} = \omega_D \frac{D_D}{D_F}$$

$$\omega_F = 300(\tfrac{1}{3}) = 100 \text{ rpm ccw} \quad \text{and} \quad \omega_F = \omega_W$$

In the same manner it can be shown that the magnitude of the tangential accelerations for X and P are

$$A_X^T = A_P^T$$

or

$$R_D \alpha_D = R_F \alpha_F$$

Solving for α_F,

$$\alpha_F = \alpha_D \frac{R_D}{R_F} = \alpha_D \frac{D_D}{D_F}$$

$$\alpha_F = 90(\tfrac{1}{3}) = 30 \text{ rad/sec}^2 \text{ ccw} \quad \text{and} \quad \alpha_F = \alpha_W$$

To determine the acceleration for point Q, use

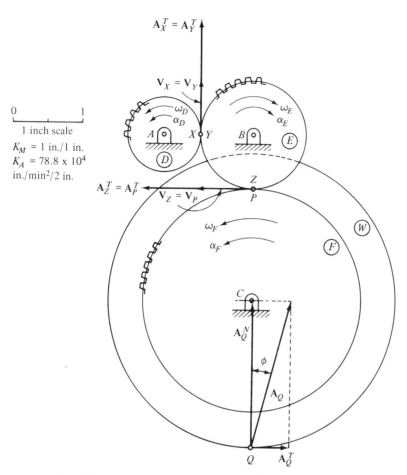

Fig. 5.8. Acceleration of Q is $A_Q = A_Q^N \leftrightarrow A_Q^T$. Only tangential components of acceleration are shown for contact points X, Y, Z, and P.

$$A_Q = A_Q^N \leftrightarrow A_Q^T$$
$$A_Q^N = R_W \omega_W^2 = 2[(100)2\pi]^2 = 78.8 \times 10^4 \text{ in./min}^2$$
$$A_Q^T = R_W \alpha_W = 2[30(3600)] = 21.6 \times 10^4 \text{ in./min}^2$$
$$A_Q = \sqrt{(78.8 \times 10^4)^2 + (21.6 \times 10^4)^2}$$
$$A_Q = 8.16 \times 10^5 \text{ in./min}^2 \quad \text{and} \quad \phi = 15.3° = \text{arc tan}\left(\frac{21.6}{78.8}\right)$$

The acceleration of Q is shown to scale in Fig. 5.8.

5.7 Linear Acceleration for
Fixed Points on a Link

It was established in Sect. 5.5 that the acceleration for point P with respect to a stationary point O, on the same link, can be resolved into normal and tangential accelerations. If point O is not stationary but has an acceleration, the same component equations are applicable; however, the acceleration of P with respect to O is now called a relative acceleration.

Let us analyze a simple example where point O is stationary. In Fig. 5.9(a), link OP is rotating and accelerating cw about connection O. The mechanism driving OP is not shown. Vehicle Ⓥ, which carries the link, is stationary. Knowing ω, α, and length OP, the acceleration of P with respect to O can be established by

$$A_{P/O} = A_{P/O}^N + A_{P/O}^T = \overline{OP}\omega^2 + \overline{OP}\alpha \qquad (1)$$

The vector addition for this is shown in Fig. 5.9(a). Since the vehicle is stationary, the absolute acceleration of P (relative to the earth) is $A_{P/O}$.

Let us modify the foregoing example by accelerating the vehicle [Fig. 5.9(b)]; label the vehicle acceleration $A_{V/E}$. Also, let the absolute angular velocity and

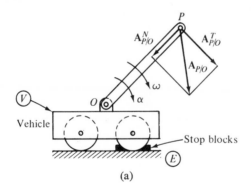

(a)

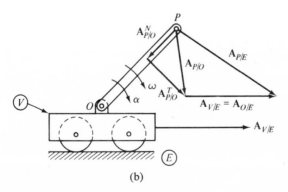

(b)

Fig. 5.9. Absolute acceleration for point P. (a) Vehicle stationary, $A_{P/E} = A_{P/O} = A_{P/O}^N + A_{P/O}^T$. (b) Vehicle accelerating, $A_{P/E} = A_{P/O}^N + A_{P/O}^T + A_{O/E}$.

acceleration for OP be exactly the same values shown in Fig. 5.9(a). Now, to determine the absolute acceleration of P requires that we consider the acceleration of the vehicle, or point O. As with linear velocity (Sect. 4.11) the points involved here can be related to one another by a relative equation. Thus the absolute acceleration for P is

$$A_{P/E} = A_{P/O} + A_{O/E} \qquad (2)$$

where $A_{P/O}$ is evaluated by Eq. (1). Since the same numerical values of ω and α are involved, the result for $A_{P/O}$ is the same as that shown in Fig. 5.9(a). The final vector equation for $A_{P/E}$, in component form, is

$$A_{P/E} = (A_{P/O}^N + A_{P/O}^T) + A_{O/E} \qquad (3)$$

since the vehicle is translating $A_{O/E} = A_{V/E}$. The graphical addition of Eq. (3) is shown in Fig. 5.9(b).

Normal acceleration can be calculated from one of the three equations

$$A_{P/O}^N = \overline{OP}\omega^2 = \frac{V_{P/O}^2}{OP} = V_{P/O}\omega$$

where ω is the absolute angular velocity for OP, and $V_{P/O}$ is the relative linear velocity of P with respect to O. Take careful note that linear velocity is a relative quantity.

5.8 Applications of the Relative Acceleration Equation

This section is concerned with the application of the relative acceleration equation to pinned connected linkages. The acceleration analysis will require a graphical layout of the equation, and, hence, it is referred to as a graphical method. A graphical analysis is rather fast and simple; however, it is subjected to drawing inaccuracies and is limited to the analysis of one position of the mechanism.

To solve the acceleration equation, it is necessary to separate accelerations into tangential and normal components. The tangential accelerations usually cannot be initially evaluated because α is unknown. Normal accelerations can be calculated from $R\omega^2$, where angular velocities are determined from a velocity analysis of the linkage. Thus an acceleration analysis is often preceded by a velocity analysis.

The acceleration analysis is explicitly illustrated in Examples 5.5 and 5.6.

EXAMPLE 5.5

Given: Figure 5.10(a); at this instant, crank Ⓒ has an angular velocity of 3000 rad/min ccw and an angular acceleration of 24,000 rad/min² ccw. Length $AB = 1$ in., and $BC = 3$ in. Point P is located midway between B and C.

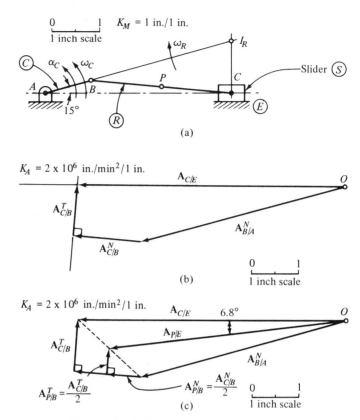

Fig. 5.10. *Acceleration analysis of a slider crank. (a) Mechanism. (b) and (c) Acceleration polygons.*

Determine: Absolute linear acceleration for the slider and point P.

Solution. Before we start the acceleration analysis it is necessary to establish the angular velocity for link $\textcircled{R}$.

STEP 1. Locate I_R; refer to Fig. 5.10(a). The intersection of extended line AB and a line perpendicular to AC through C locates the instant center I_R. From the mechanism layout, $BI_R = 3.09$ in.

STEP 2. Determine ω_R:

$$V_B = BA\omega_C \quad \text{and} \quad V_B = BI_R\omega_R$$

Equate and solve for ω_R:

$$\omega_R = \frac{BA}{BI_R}\omega_C = \frac{1}{3.09}(3000) = 971 \frac{\text{rad}}{\text{min}} \text{ cw}$$

STEP 3. The relative acceleration equation relating to points C and B is

$$A_{C/E} = A_{C/B} +\!\!\!\!+ A_{B/E}$$

where

$$A_{C/B} = A_{C/B}^N +\!\!\!\!+ A_{C/B}^T$$

and

$$A_{B/E} = A_{B/A} = A_{B/A}^N +\!\!\!\!+ A_{B/A}^T$$

Combining components into one equation,

$$A_{C/E} = (A_{C/B}^N +\!\!\!\!+ A_{C/B}^T) +\!\!\!\!+ (A_{B/A}^N +\!\!\!\!+ A_{B/A}^T)$$

Calculate the normal and tangential components, if possible:

$A_{B/A}^N = BA\omega_C^2 = 1(3000)^2 = 9 \times 10^6$ in./min²   (parallel to BA)

$A_{B/A}^T = BA\alpha_C = 1(24,000) = 24 \times 10^3$ in./min²   (perpendicular to BA)

$A_{C/B}^N = CB\omega_R^2 = 3(971)^2 = 2.83 \times 10^6$ in./min²   (parallel to CB)

$A_{C/B}^T = CB\alpha_R = 3(\alpha_R) = ?$ in./min²   (perpendicular to CB)

The magnitude and vector sense for $A_{C/B}^T$ are unknown; however, the line of action is known, it is perpendicular to CB.

STEP 4. To solve the vector equation, place the unknown terms at the end of the vector addition. Rearranging, the equation now reads as

$$A_{C/E} = (A_{B/A}^N +\!\!\!\!+ A_{B/A}^T) +\!\!\!\!+ (A_{C/B}^N +\!\!\!\!+ A_{C/B}^T)$$

Assume an acceleration scale of $K_A = 2 \times 10^6$ in./min²/1 in. Start vector addition at point O [Fig. 5.10(b)]; use calculated data from step 3. Add $A_{C/B}^N$ to $A_{B/A}^N$. Neglect $A_{B/A}^T$; it is comparatively small. Determine $A_{C/B}^T$ and $A_{C/E}$; construct a line which passes through the head of $A_{C/B}^N$ and which is perpendicular to this vector. Also, construct a line through O, which is parallel to AC. The intersection of these lines establishes the magnitude and sense of the unknown acceleration vectors.

STEP 5. Scale off $A_{C/E}$. The slider acceleration is $A_{C/E} = 11.34 \times 10^6$ in./min². The acceleration direction is toward the left.

STEP 6. Determine $A_{P/E}$. The relative equation is

$$A_{P/E} = A_{P/B} +\!\!\!\!+ A_{B/E}$$

Change into component form and rearrange, recalling that $A_{B/E} = A_{B/A}$:

$$A_{P/E} = (A_{B/A}^N +\!\!\!\!+ A_{B/A}^T) +\!\!\!\!+ (A_{P/B}^N +\!\!\!\!+ A_{P/B}^T)$$

where

$$A_{P/B}^N = PB\omega_R^2 = \left(\frac{CB}{2}\right)\omega_R^2 = \frac{A_{C/B}^N}{2}$$

and

$$A_{P/B}^T = PB\alpha_R = \left(\frac{CB}{2}\right)\alpha_R = \frac{A_{C/B}^T}{2}$$

The components of P relative to B can be easily laid out by scaling off half of $A_{C/B}^N$ and half of $A_{C/B}^T$. The completed polygon is shown in Fig. 5.10(c). Note, again, in the vector addition, neglect $A_{B/A}^T$.

STEP 7. Scale off $A_{P/E}$. $A_{P/E} = 10.1 \times 10^6$ in./min² at an angle of 6.8 deg with the horizontal.

EXAMPLE 5.6

Given: Crank ⓒ of a geared four-bar linkage $ABCD$ [Fig. 5.11(a)] rotates at a constant 500 rpm cw. Link BC is pinned to gear ⓖ at point P. The dimensions are

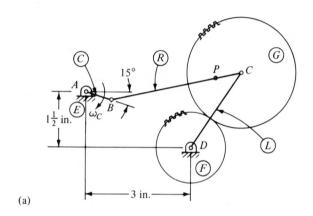

(a)

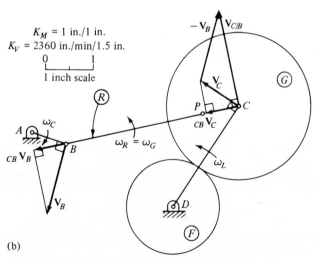

(b)

Fig. 5.11. *Acceleration analysis of a geared four-bar linkage.*
(a) Mechanism. (b) Relative velocity.

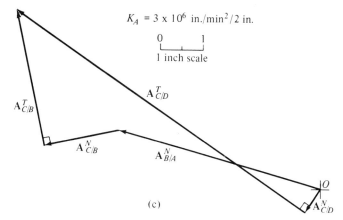

$K_A = 3 \times 10^6$ in./min²/2 in.

0 ____ 1
1 inch scale

$A_{C/D}^T$

$A_{C/B}^T$

$A_{C/B}^N$

$A_{B/A}^N$

O

$A_{C/D}^N$

(c)

Fig. 5.11. *(Cont.)* *(c) Acceleration polygon.*

$AB = \frac{3}{4}$ in., $BC = 3\frac{3}{4}$ in., $CD = 2\frac{1}{2}$ in., the diameter of gear ⓖ $= 3$ in., and the diameter of gear ⒡ $= 2$ in.

Determine: The angular acceleration for gear ⓖ.

Solution. The relative velocity method, discussed in Sect. 4.14, will be used to determine ω_R. This method is especially useful when the instant center is located far off the mechanism.

STEP 1. Determine V_C by components [refer to Fig. 5.11(b)]:

$$V_B = AB\omega_C = \tfrac{3}{4}(500 \times 2\pi) = 2360 \text{ in./min}$$

Let $K_V = 2360$ in./min/1.5 in. By construction,

$$V_C = 1540 \text{ in./min}$$

STEP 2. Determine $V_{C/B}$ using the relative equation

$$V_{C/B} = V_C \longrightarrow V_B = V_C + (-V_B)$$
$$V_{C/B} = 3210 \text{ in./min}$$

STEP 3. Calculate ω_R:

$$\omega_R = \frac{V_{C/B}}{CB} = \frac{3210}{3.75} = 857 \frac{\text{rad}}{\text{min}} \text{ccw}$$

Since CB or ⓡ is rigidly fastened to ⓖ, $\omega_R = \omega_G$.
STEP 4. Calculate ω_L:

$$\omega_L = \frac{V_C}{CD} = \frac{1540}{2.5} = 616 \frac{\text{rad}}{\text{min}} \text{ccw}$$

STEP 5. The relative acceleration equation is

$$A_{C/E} = A_{C/B} + A_{B/E}$$

161

where $A_{C/E} = A_{C/D}$ and $A_{B/E} = A_{B/A}$. The component form is

$$A_{C/D}^N \dotplus A_{C/D}^T = (A_{C/B}^N \dotplus A_{C/B}^T) \dotplus (A_{B/A}^N \dotplus A_{B/A}^T)$$

To solve graphically, change the order of addition to

$$A_{C/D}^N \dotplus A_{C/D}^T = (A_{B/A}^N \dotplus A_{B/A}^T) \dotplus (A_{C/B}^N \dotplus A_{C/B}^T)$$

Evaluate all components, if possible:

$A_{C/D}^N = CD\omega_L^2 = 2.5(616)^2 = .95 \times 10^6 \text{ in./min}^2$ (parallel to CD)

$A_{C/D}^T = CD\alpha_L = 2.5(\alpha_L) = ? \text{ in./min}^2$ (perpendicular to CD)

$A_{B/A}^N = BA\omega_C^2 = .75(3140)^2 = 7.4 \times 10^6 \text{ in./min}^2$ (parallel to BA)

$A_{B/A}^T = BA\alpha_C = .75(0) = 0$

$A_{C/B}^N = CB\omega_R^2 = 3.75(857)^2 = 2.75 \times 10^6 \text{ in./min}^2$ (parallel to CB)

$A_{C/B}^T = CB\alpha_R = 3.75(\alpha_R) = ? \text{ in./min}^2$ (perpendicular to CB)

STEP 6. Solve for $A_{C/B}^T$ by the acceleration polygon [Fig. 5.11(c)]. Use the acceleration scale $K_A = 3 \times 10^6$ in./min²/2 in. Starting at O, add $A_{C/B}^N$ to $A_{B/A}^N$. Place the tail of the resultant component $A_{C/D}^N$ at point O. To obtain $A_{C/B}^T$ and $A_{C/D}^T$, construct a line perpendicular to $A_{C/D}^N$ and another line perpendicular to $A_{C/B}^N$. Vector heads are drawn together at the intersection of these lines.

STEP 7. Scale off $A_{C/B}^T$. $A_{C/B}^T = 4.8 \times 10^6$ in./min².

STEP 8. Calculate the angular acceleration for Ⓖ:

$$\alpha_G = \alpha_R = \frac{A_{C/B}^T}{CB} = \frac{4.8 \times 10^6}{3.75} = 1.28 \times 10^6 \frac{\text{rad}}{\text{min}^2} \text{ccw}$$

To establish the direction of rotation for α_R, place vector $A_{C/B}^T$ at point C. The sense of $A_{C/B}^T$ indicates that α_R is ccw.

5.9 Accelerations for Bodies in Pure Rolling

The contact points for bodies in pure rolling or no slip generally have different accelerations; however, the component accelerations in the tangential direction are the same, see Sect. 5.6.

Figure 5.12(a) shows three gears in pure rolling; here gear Ⓖ is driven by rack Ⓜ over a stationary rack Ⓡ. Rack Ⓜ is translating at a constant linear velocity, and gear Ⓖ is revolving at a constant angular velocity.

The acceleration for point P is determined by

$$A_P = A_{P/O} \dotplus A_O \tag{1}$$

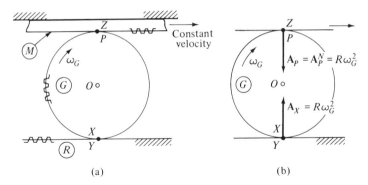

Fig. 5.12. *Pure rolling, rack (M) moving at constant velocity and gear (G) revolving at constant angular velocity.*

or, in component form,

$$A_P^N +\!\!\!> A_P^T = A_O +\!\!\!> A_{P/O}^N +\!\!\!> A_{P/O}^T \qquad (2)$$

Note that the subsymbol E, representing the earth, has been dropped simply for writing convenience. Gear (G) is not accelerating; thus, $A_O = 0$ and $A_{P/O}^T = R\alpha_G = 0$. Since A_P^T is the only remaining tangential component in the component equation [Eq. (2)], $A_P^T = 0$. The component equation simplifies to $A_P^N = A_{P/O}^N$. Since $A_{P/O}^N = R\omega_G^2$; Eq. (1) simplifies to $A_P = R\omega_G^2$, where A_P is directed toward O [see Fig. 5.12(b)]. Note that for pure rolling the components for A_P and A_Z in the horizontal direction are the same, namely, zero.

Let us now establish the acceleration for contact point X:

$$A_X = A_{X/O} +\!\!\!> A_O$$

or

$$A_X^N +\!\!\!> A_X^T = A_O +\!\!\!> A_{X/O}^N +\!\!\!> A_{X/O}^T$$

In the same manner as previously indicated, $A_O = 0$, $A_{X/O}^T = 0$, and $A_X^T = 0$; thus, $A_X = R\omega_G^2$, where A_X is directed upward toward O. Note that point X is the instant center for gear (G). Here is a good example where a point, at this instant, has a zero velocity as well as an acceleration. Being that (G) is in pure rolling with (R) the components of A_X and A_Y in the horizontal direction are equal, both components are zero.

In Fig. 5.13(a), rack (M) is accelerating with an acceleration of A_Z. The component of A_P in the horizontal direction is A_Z; that is, $A_P^T = A_Z$. Since point O is moving horizontally at half the velocity of P, it can be concluded that $A_O = \frac{1}{2}(A_P^T)$ or $2A_O = A_P^T$. The acceleration for point P can be obtained from

$$A_P = A_{P/O} +\!\!\!> A_O$$

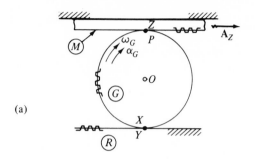

(a)

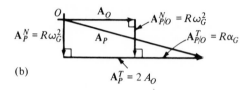

(b)

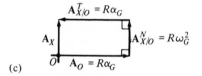

(c)

Fig. 5.13. Pure rolling, rack Ⓜ accelerating and gear Ⓖ has an angular acceleration.

or

$$A_P^N + A_P^T = A_O + A_{P/O}^N + A_{P/O}^T$$

Although the numerical values for the accelerations are unknown, the acceleration vectors can be laid out in the following sequence. Refer to Fig. 5.13(b). Starting at O, first lay out A_O; then lay out $A_{P/O}^N = R\omega_G^2$ perpendicular to A_O; next lay out the line of action for $A_{P/O}^T = R\alpha_G$, which is perpendicular to $A_{P/O}^N$. The vector sense for $A_{P/O}^T$ is toward the right because α_G is cw. Now, from point O lay out A_P^N, recalling from Eq. (2) that $A_P^N = R\omega_G^2$. Note that $A_P^N = A_{P/O}^N = R\omega_G^2$. To complete the polygon, lay out the resultant component A_P^T perpendicular to A_P^N, where $A_P^T = 2A_O$. From the acceleration polygon it can be seen that $A_O = A_{P/O}^T$; thus, $A_O = R\alpha_G$. The magnitude of the resultant acceleration A_P is

$$A_P = \sqrt{(R\omega_G^2)^2 + (2R\alpha_G)^2}$$

Let us now determine the acceleration for instant center X. Using the equation

$$A_X = A_O + A_{X/O}^N + A_{X/O}^T$$

refer to Fig. 5.13(c), add $A_O = R\alpha_G$, $A_{X/O}^N = R\omega_G^2$, and $A_{X/O}^T = R\alpha_G$. The vector resultant is $A_X = R\omega_G^2$, and it is directed upward toward point O. Note that A_X is the same value obtained when Ⓖ was rotating at constant angular velocity.

In summary, for a rolling gear or disk on a stationary flat surface, the center point has an acceleration of $R\alpha_G$, and the instant center has an acceleration of $R\omega_G^2$ directed toward the center.

EXAMPLE 5.7

Given: The center O of a disk (Fig. 5.14) is moving to the left at 50 ft/sec and decelerating at 20 ft/sec². The disk diameter is 6 in.

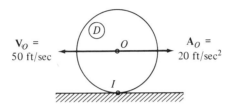

$V_O = $ 50 ft/sec $A_O = $ 20 ft/sec²

Fig. 5.14. Disk in pure rolling.

Determine: Angular velocity and angular acceleration for the disk; assume pure rolling.

Solution. The instant center I for the disk is at the contact point with the ground

$$\omega_D = \frac{V_O}{OI} = \frac{50}{\frac{3}{12}} = 200 \frac{\text{rad}}{\text{sec}} \text{ ccw}$$

To determine angular acceleration, use

$$A_O = R\alpha_D, \qquad \alpha_D = \frac{A_O}{R} = \frac{20}{\frac{3}{12}} = 80 \frac{\text{rad}}{\text{sec}^2} \text{ cw}$$

5.10 Coriolis Acceleration

The relative acceleration equation, established in Sect. 5.7, relates to the acceleration of two points on the same body. If the two points are adjacent or coincident points on different rotating bodies, the relative acceleration equation, as it is now understood, is not applicable.

Figure 5.15(a) shows a simplified sketch of a cylinder Ⓛ rotating cw at a constant angular velocity about point A. At this same instant, the piston rod Ⓡ is moving outwardly with a constant linear velocity relative to the cylinder. It is desirable to establish the absolute acceleration for point Y on rod Ⓡ. The relative acceleration equation used previously does not fit this mechanism because distance YA is not constant but is always changing. Let us attempt to solve for A_Y by using the following procedure:

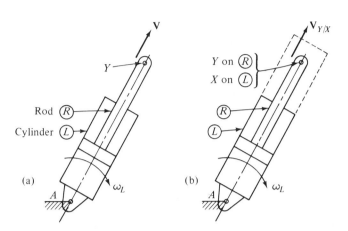

Fig. 5.15. The cylinder rotates cw as the piston rod moves outwardly relative to the cylinder.

1. Establish a point X on L which is coincident with point Y. This is accomplished by extending L [Fig. 5.15(b)].
2. Determine the acceleration of Y relative to X. To help determine Y relative to X, imagine L to be held stationary.
3. Determine the absolute acceleration of point X.

Expressing steps 2 and 3 in equation form, we obtain

$$A_Y = A_{Y/X} + A_X$$

Although the sequence seems to be logical, the equation obtained is not complete. What is missing is a term called the Coriolis component of acceleration. Equation (5.5) shows the complete acceleration equation which includes the Coriolis acceleration A_{cor}:

$$A_Y = A_{Y/X} + A_X + A_{cor} \tag{5.5}$$

The component form of this equation is

$$A_Y^N + A_Y^T = A_{Y/X}^N + A_{Y/X}^T + A_X^N + A_X^T + A_{cor} \tag{5.6}$$

Equations (5.5) and (5.6) can become unwieldy; therefore it is suggested that the following designations be used. Label coincident points Y and X, where Y is on R and X is on the rotating link L. Be careful in choosing point Y. Point Y must trace a straight line or a circular arc relative to L. The center of curvature for Y, relative to L, will be labeled O. If Y traces out a straight line on L, O is located at infinity.

The following is a description of each term in Eq. (5.6):

A_Y^N and A_Y^T are the absolute component accelerations for point Y on link $\circledR$.

$A_{Y/X}^N$ and $A_{Y/X}^T$ are the component accelerations of point Y relative to point X. $A_{Y/X}^N = YO\omega_{R/L}^2 = V_{Y/X}^2/YO = V_{Y/X}\omega_{R/L}$, where the relative quantities $\omega_{R/L} = \omega_R - \omega_L$ and $V_{Y/X} = V_Y \rightarrow V_X$. The normal acceleration is directed toward O, along line YO. The tangential component $A_{Y/X}^T = YO\alpha_{R/L}$ acts perpendicular to YO in a direction consistent with $\alpha_{R/L}$. Note that $\alpha_{R/L} = \alpha_R - \alpha_L$. If the trace of Y relative to $\circledL$ is a straight line, as in translation, $\alpha_{R/L} = 0$; however $A_{Y/X}^T$ may not be zero.

A_X^N and A_X^T are the absolute component accelerations for point X on link $\circledL$.

A_{cor} symbolizes the Coriolis acceleration; the magnitude of $A_{cor} = 2V_{Y/X}\omega_L$. To determine the direction for A_{cor}, rotate the relative velocity vector $V_{Y/X}$ about its tail, 90 deg in the same direction as ω_L. Thus A_{cor} is perpendicular to $V_{Y/X}$. Figure 5.16 shows four possible situations for A_{cor}. The Coriolis acceleration

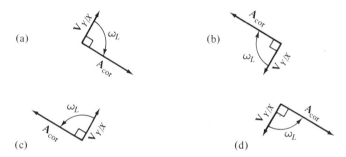

Fig. 5.16. The direction for the Coriolis vector is determined by pivoting $V_{Y/X}$ 90 deg in the same sense as ω_L.

is due to a change in the direction of the relative velocity $(V_{Y/X})$; it is also affected by the tangential velocity $(R\omega_L)$ continually changing as the radius changes.

It is important that we recognize a Coriolis-type problem. If a point moves on a known path, which has rotation, the Coriolis component is applicable to this problem. Again, it is emphasized that for our purposes the path must be either a straight line or a circular arc.

To illustrate a numerical problem involving the Coriolis component, refer to Fig. 5.17. For the instant shown, cylinder $\circledL$ has a constant angular velocity of 20 rad/sec cw, piston rod $\circledR$ has a constant outward velocity of 24 in./sec, and distance $XA = 8$ in. Determine the acceleration of Y. Note that Y traces a straight line relative to $\circledL$. Equation (5.6), slightly modified, reads

$$A_Y = A_{Y/X}^N + A_{Y/X}^T + A_X^N + A_X^T + A_{cor}$$

$$A_Y = ?$$

$$A_{Y/X}^N = \frac{V_{Y/X}^2}{YO} = 0, \qquad \text{path on } \circledL \text{ is straight; thus, } YO = \infty$$

$$A_{Y/X}^T = 0, \qquad V_{Y/X} \text{ is constant}$$

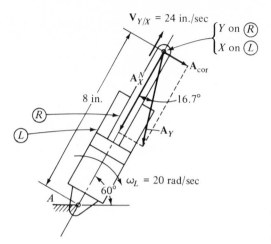

Fig. 5.17. Rotating cylinder requires the use of the Coriolis acceleration.

$$A_X^N = XA\omega_L^2 = 8(20)^2 = 3200 \text{ in./sec}^2 \quad \text{(parallel to } XA)$$

$$A_X^T = XA\alpha_L = 0, \quad \alpha_L = 0$$

$$A_{cor} = 2V_{Y/X}\omega_L = 2(24)(20) = 960 \text{ in./sec}^2 \quad \text{(perpendicular to } V_{Y/X})$$

The equation reduces to

$$A_Y = A_X^N + A_{cor} = 3200 + 960 = 3340 \text{ in./sec}^2$$

at an angle of 16.7 deg with A_X^N; refer to Fig. 5.17.

EXAMPLE 5.8

Given: Figure 5.18(a); an input crank rotates cw about B at 40 rad/sec and accelerates at 400 rad/sec².

Determine: The linear acceleration for a fictitious point Q, where Q is on the horizontal link.

Solution. Let us first identify point Y. Relative to the horizontal link the center of the slider traces a straight line on ⓛ. Hence, Y is on the input crank ⓡ and X is on the horizontal link ⓛ. Refer to the kinematic sketch shown in Fig. 5.18(b).

STEP 1. Determine V_X by the component method. Refer to Fig. 5.18(b). First, calculate V_Y:

$$V_Y = YB\omega_R = \frac{6.4}{12}(40) = 21.3 \frac{\text{ft}}{\text{sec}}$$

Let $K_V = 21.3$ ft/sec/2 in. By construction, $V_X = 13.7$ ft/sec.

STEP 2. Determine $V_{Y/X}$. Use the relative equation

$$V_{Y/X} = V_Y \longrightarrow V_X$$

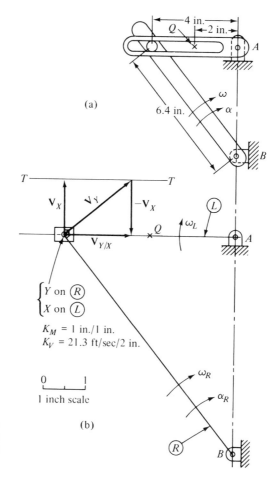

Fig. 5.18. Acceleration analysis. (a) Mechanism sketch. (b) Kinematic drawing and vector diagram.

The vector construction is shown in Fig. 5.18(b):

$$V_{Y/X} = 16.8 \text{ ft/sec}$$

STEP 3. Calculate ω_L:

$$\omega_L = \frac{V_X}{XA} = \frac{13.7}{\frac{4}{12}} = 41.1 \frac{\text{rad}}{\text{sec}} \text{ cw}$$

STEP 4. Solve for A_X; use Eq. (5.6):

$$A_Y^N + A_Y^T = A_{Y/X}^N + A_{Y/X}^T + A_X^N + A_X^T + A_{\text{cor}}$$

$$A_Y^N = YB\omega_R^2 = \frac{6.4}{12}(40)^2 = 853 \frac{\text{ft}}{\text{sec}^2} \qquad \text{(parallel to } YB)$$

$$A_Y^T = YB\alpha_R = \frac{6.4}{12}(400) = 213 \frac{\text{ft}}{\text{sec}^2} \qquad \text{(perpendicular to } YB)$$

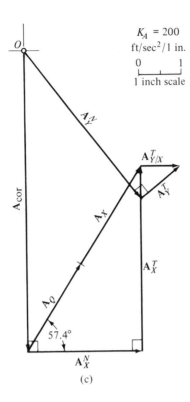

$$K_A = 200 \text{ ft/sec}^2/1 \text{ in.}$$

1 inch scale

(c)

Fig. 5.18. (Cont.) (c) Acceleration polygon.

$$A^N_{Y/X} = \frac{V^2_{Y/X}}{YO} = 0, \qquad \text{path on } \textcircled{L} \text{ is straight; thus, } YO = \infty$$

$$A^T_{Y/X} = ? \qquad (\text{parallel to } V_{Y/X}) \longleftrightarrow$$

$$A^N_X = \frac{V^2_X}{XA} = \frac{(13.7)^2}{\frac{4}{12}} = 539 \frac{\text{ft}}{\text{sec}^2} \qquad (\text{parallel to } XA) \longrightarrow$$

$$A^T_X = XA\alpha_L = ? \qquad (\text{perpendicular to } XA) \Big\uparrow$$

$$A_{\text{cor}} = 2V_{Y/X}\omega_L = 2(16.8)(41.1) = 1380 \text{ ft/sec}^2$$
$$(\text{perpendicular to } V_{Y/X}) \Big\downarrow$$

STEP 5. The order of solution for the acceleration polygon [Fig. 5.18(c)] is $A^N_Y + A^T_Y = A_{\text{cor}} + A^N_X + A^T_X + A^T_{Y/X}$. Let $K_A = 200$ ft/sec^2/1 in. Starting at O, add A^N_Y and A^T_Y. Again starting at O, add A_{cor} and A^N_X. A^T_X and $A^T_{Y/X}$ are unknown. Draw the line of action for A^T_X through the head of vector A^N_X. Also, draw the line of action for $A^T_{Y/X}$ through the head of A^T_Y. The intersection of these lines establishes A^T_X and $A^T_{Y/X}$.

STEP 6. A_X is determined by the vector addition of A^N_X and A^T_X:

$$A_X = 1000 \text{ ft/sec}^2$$

STEP 7. Since $QA = \frac{1}{2}(XA)$, it follows that $A_Q = \frac{1}{2}A_X$; thus, $A_Q = 500$ ft/sec^2 at an angle of 57.4 deg with XA.

EXAMPLE 5.9

Given: A dual centrifuge [Fig. 5.19(a)] consists of a large turntable onto which is mounted, at position B, a small satellite turntable. The large table rotates about A at 1000 rpm cw. The satellite turntable is independently driven by a nonslip belt at 500 rpm cw (relative to the large table).

Determine: The absolute acceleration for a point on the satellite table which is 28 in. from A.

Solution. This problem can be solved without using the Coriolis component. To compare each method, both solutions will be shown. The first is without the Coriolis component.

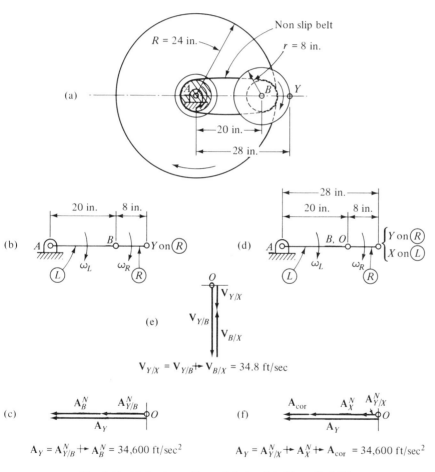

$$V_{Y/X} = V_{Y/B} + V_{B/X} = 34.8 \text{ ft/sec}$$

$$A_Y = A^N_{Y/B} + A^N_B = 34,600 \text{ ft/sec}^2$$

$$A_Y = A^N_{Y/X} + A^N_X + A_{cor} = 34,600 \text{ ft/sec}^2$$

Fig. 5.19. (a) Dual centrifuge. Solution 1 (without Coriolis): (b) kinematic drawing, (c) acceleration diagram. Solution 2 (with Coriolis): (d) kinematic drawing, (e) velocity diagram, (f) acceleration diagram.

Solution 1 (without Coriolis)

The kinematic drawing is shown in Fig. 5.19(b). The large turntable is labeled $\textcircled{L}$, the satellite table is marked $\textcircled{R}$, and the outside point is labeled Y.

STEP 1. Determine ω_R. The angular velocity of $\textcircled{R}$ relative to $\textcircled{L}$ was given as 500 rpm cw and $\omega_L = 1000$ rpm cw. Assume cw to be plus; thus, $\omega_R = \omega_{R/L} + \omega_L = 500 + 1000 = 1500$ rpm cw.

STEP 2. Convert ω_R and ω_L into radians per second:

$$\omega_R = 1500\left(\frac{2\pi}{60}\right) = 157\frac{\text{rad}}{\text{sec}}\text{ cw}$$

$$\omega_L = 1000\left(\frac{2\pi}{60}\right) = 104.5\frac{\text{rad}}{\text{sec}}\text{ cw}$$

STEP 3. The relative acceleration equation in component form is

$$A_Y = A^N_{Y/B} + A^T_{Y/B} + A^N_B + A^T_B$$

$A_Y = ?$

$A^N_{Y/B} = YB\omega^2_R = \frac{8}{12}(157)^2 = 16400 \text{ ft/sec}^2 \qquad$ (parallel to YB) ◄————

$A^T_{Y/B} = YB\alpha_R = 0, \qquad \alpha_R = 0$

$A^N_B = BA\omega^2_L = \frac{20}{12}(104.5)^2 = 18200 \text{ ft/sec}^2 \qquad$ (parallel to BA) ◄————

$A^T_B = BA\alpha_L = 0, \qquad \alpha_L = 0$

STEP 4. Solve the vector equation for A_Y [Fig. 5.19(c)]. $A^N_{Y/B}$ and A^N_B are collinear; therefore, by simple addition, $A_Y = 34,600 \text{ ft/sec}^2$ horizontal.

Solution 2 (with Coriolis)

The kinematic drawing has been slightly revised to include point X on the large turntable $\textcircled{L}$ [see Fig. 5.19(d)]. Relative to $\textcircled{L}$ point Y traces a circular path on $\textcircled{L}$. Thus the center of curvature for Y is at B; label this point O.

STEP 1. Determine $V_{Y/X}$. Use the relative equation

$$V_{Y/X} = V_{Y/B} + V_{B/X}$$

$$V_{Y/B} = YB\omega_R = \frac{8}{12}(157) = 104.5 \text{ ft/sec}$$

$$V_{B/X} = BX\omega_L = \frac{8}{12}(104.5) = 69.7 \text{ ft/sec}$$

The vector addition is shown in Fig. 5.19(e): $V_{Y/X} = 34.8 \text{ ft/sec}$

STEP 2. To solve for A_Y, use Eq. (5.6):

$$A_Y = A^N_{Y/X} + A^T_{Y/X} + A^N_X + A^T_X + A_{\text{cor}}$$

$A_Y = ?$

$$A_{Y/X}^N = \frac{V_{Y/X}^2}{YO} = \frac{(34.8)^2}{\frac{8}{12}} = 1820 \frac{ft}{sec^2} \qquad \text{(parallel to } YO\text{)} \longleftarrow$$

$$A_{Y/X}^T = YO\alpha_{R/L} = 0, \qquad \alpha_{R/L} = 0$$

$$A_X^N = XA\omega_L^2 = \tfrac{28}{12}(104.5)^2 = 25{,}500 \text{ ft/sec}^2 \qquad \text{(parallel to } XA\text{)} \longleftarrow$$

$$A_X^T = XA\alpha_L = 0, \qquad \alpha_L = 0$$

$$A_{cor} = 2V_{Y/X}\omega_L = 2(34.8)(104.5) = 7280 \text{ ft/sec}^2 \qquad \text{(perpendicular to } V_{Y/X}\text{)} \longleftarrow$$

STEP 3. The vector solution is shown in Fig. 5.19(f): $A_Y = 34{,}600$ ft/sec^2 horizontal.

A comparison between solution 1 and solution 2 shows that we obtained the same answer. The non-Coriolis solution is, of course, simpler.

PROBLEMS

5.1 A gear starts from rest and accelerates at a constant 50 rad/sec^2 about its center. Determine the instantaneous angular velocity and the number of revolutions it has rotated after a 2-sec interval.

5.2 A rotating link increases its angular velocity from 1000 to 1200 rpm, with constant angular acceleration, in .25 sec. Determine the angular acceleration in rad/sec^2.

5.3 Crank AB of slider crank ABC starts from a 30-deg position, as shown in Fig. P5.3, and accelerates at a constant 10π rad/sec^2 cw for 7 sec. What is the velocity of the slider 4 sec after start? Crank $AB = 1.5$ in., and $BC = 3$ in.

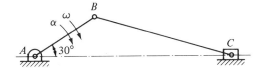

Fig. P5.3.

5.4 A gear is acted upon by a 12 in.-lb torque. The mass moment of inertia is .002 slug-ft^2. Determine the angular acceleration from the equation $T = I\alpha$, where $T = $ torque (ft-lb), $I = $ mass moment of inertia (slug-ft^2), and $\alpha = $ angular acceleration (rad/sec^2).

5.5 A car traveling at 60 mph comes to a complete stop in 15 sec. What is the deceleration in ft/sec^2 (assume constant deceleration) and the distance traveled in this time period? Sketch a linear velocity-time diagram.

5.6 Figure P5.6 shows a large centrifuge used to test human reactions to high centrifugal forces. If the rotating arm has a 50-ft radius and is rotating at 25 rpm, what is the normal acceleration? Give your answer in g's, where 1 g is 32.2 ft/sec^2.

Fig. P5.6. (Courtesy of NASA.)

5.7 Link AB of a four-bar linkage $ABCD$ (Fig. P5.7) rotates at a constant angular velocity of 50 rad/sec ccw. Determine, for the position shown, the normal acceleration of point E relative to D. CED is one link. $AB = .75$ in., $BC = 2.5$ in., $CD = ED = 2.0$ in., and $DA = 3.5$ in.

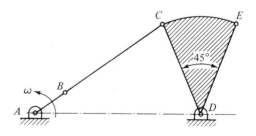

Fig. P5.7.

5.8 Starting from the position shown in Fig. P5.7, rotate link AB 60 deg ccw. Using 50 rad/sec, determine the normal acceleration for point E relative to D.

5.9 Determine the inertia force for a shaft rotating at 8000 rpm which has a dynamic unbalance of 1 oz-in. A 1 oz-in. unbalance can be interpreted as 1 oz located 1 in. from the axis of rotation. Use the equation $F = (W/g)A^N$, where $F =$ inertia force (lb), $W =$ weight (lb), $g =$ gravity acceleration (32.2 ft/sec²), and $A^N =$ normal acceleration (ft/sec²). Normal acceleration is calculated for a point 1 in. from the axis of rotation.

5.10 At a particular instant, a 4-in. diameter eccentric cam, with an eccentricity of 1 in., rotates at 250 rpm ccw with a deceleration of 250 rad/sec². What is the maximum linear acceleration for a point on the periphery of the cam?

5.11 Gear $Ⓐ$ (Fig. P5.11) has an angular acceleration of $\frac{1}{2}$ rad/sec². Determine the angular acceleration for pulley $Ⓓ$. Assume no slip. The diameters are $Ⓐ =$ .5 in., $Ⓑ = 1.0$ in., $Ⓒ = 1.25$ in., and $Ⓓ = 1.5$ in.

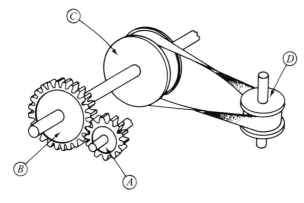

Fig. P5.11.

5.12 The disk $Ⓓ$ of the disk and wheel transmission shown in Fig. P5.12 has an angular velocity of 5 rad/sec and an angular acceleration of 15 rad/sec². Assuming no slip, determine the linear acceleration for the contact point on the wheel.

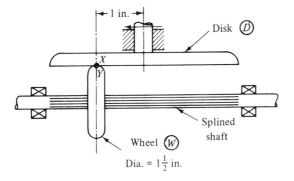

Fig. P5.12.

5.13 In Fig. P5.13, a rack starts from rest and moves at a constant acceleration for 3 sec. Determine the tangential acceleration for point P on gear $\circledR$ and the angular acceleration for gear $\circledR$ at 2 sec from rest. The rack has attained a linear velocity of 4 in./sec at this time. The diameters for $\circledA$ and $\circledR$ are 1 and 2 in., respectively.

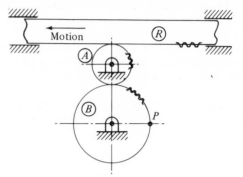

Fig. P5.13.

5.14 A bicycle starts from rest and travels 100 ft in 10 sec, with constant acceleration along a level road. Determine (a) the linear acceleration of the bike, (b) the angular acceleration of the foot pedal arms, and (c) the angular velocity of the foot pedal arms at the end of 10 sec. The back wheel chain sprocket and foot pedal sprocket diameters are 3 and 7 in., respectively. The wheel diameters are 20 in. The linear acceleration for the wheel center can be expressed by the equation $A = R\alpha$, where R is the wheel radius and α is the angular acceleration of the wheel. Assume a condition of pure rolling.

5.15 A two-cycle gasoline engine operates at 3000 rpm. The crank length is 8.750 in., and the connecting rod length is 3.125 in. Determine the piston acceleration when the crank is 30 deg from TDC (piston approaching TDC). Is the piston decelerating?

5.16 For the position shown in Fig. P5.16, determine the angular acceleration for gear $\circledF$. Link AB rotates at 200 rpm cw. $AB = 1$ in., $BC = 5$ in., and the offset $= 2\frac{1}{2}$ in.

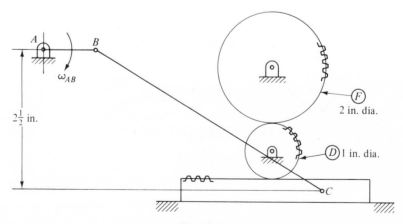

Fig. P5.16.

5.17 Using the dimensions and rpm shown in Fig. P5.16, determine the angular acceleration and angular velocity for gear Ⓕ when point C is at midstroke (the rack is moving toward the right).

5.18 For the linkage position shown in Fig. P5.18, determine the acceleration for point B. Crank CD rotates at 50 rad/sec cw. $CD = 2$ in., $CB = 6$ in., $AB = 3$ in., and $DA = 5.5$ in.

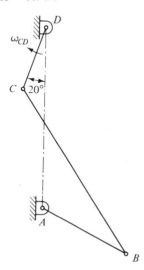

Fig. P5.18.

5.19 Using the dimensions and rpm shown in Fig. P5.18, determine the acceleration for point B and the angular acceleration for AB when point B is at its uppermost position.

5.20 Figure P5.20 shows a slider crank. For the 30-deg position shown, determine the acceleration of the slider using the following equations. Link lengths will be in inches and angular velocities are in radians per second. Crank Ⓒ rotates 1000 rad/sec ccw. $AB = 2$ in., and $BC = 5$ in. If the quantity calculated is negative, retain the sign in subsequent calculations.

$$\omega_L = \left[\frac{(AB/BC)\cos\theta}{\sqrt{1-(AB/BC)^2 \sin^2\theta}} \right] \omega_C$$

$$\alpha_L = \frac{\omega_C^2(AB/BC)\sin\theta[(AB/BC)^2 - 1]}{[1-(AB/BC)^2 \sin^2\theta]^{3/2}}$$

$$A_C = -\omega_C^2(AB)\left(\cos\theta + \alpha_L \frac{\sin\theta}{\omega_C^2} + \omega_L \frac{\cos\theta}{\omega_C}\right)$$

Fig. P5.20.

5.21 A slider crank is attached to a vehicle (Fig. P5.21) which is accelerating at 1000 ft/sec². Determine the absolute linear acceleration for the slider when the slider is at top dead center. AB rotates at 2000 rpm cw. $AB = .5$ in., and $BC = 2$ in.

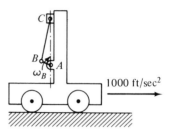

Fig. P5.21.

5.22 At this instant, the slider in Fig. P5.22 has an upward velocity of 1 ft/sec with an acceleration of 20 ft/sec². Determine the acceleration for points E and P. $AB = 1.75$ in., $CB = 1.0$ in., $CBD = 2.0$ in., $CF = 4.0$ in., $EF = 3.0$ in., $DE = 3.0$ in., and $FP = 1.0$ in.

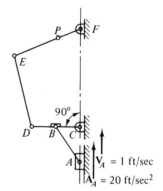

Fig. P5.22.

5.23 Four-bar linkage $ABCD$ (Fig. P5.23) is driven by AB at 1 rad/sec ccw. For the position shown, determine the radius of curvature for point P. Use the equation $\rho = V_P^2/A_P^N$, where ρ = radius of curvature, V_P = absolute linear velocity of P, and A_P^N = absolute normal acceleration of P. The normal acceleration vector is perpendicular to V_P and is directed toward the center of curvature. To establish ρ, first determine A_C; then determine A_P, then construct A_P^N, and then finally calculate ρ. The dimensions are $AB = 3.25$ in., $BC = 3.25$ in., $CD = 2.75$ in., $AD = 6.50$ in., and $BP = 1.25$ in.

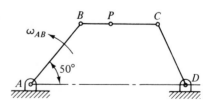

Fig. P5.23.

5.24 A 2-in. diameter gear rolls on a stationary rack. If the angular velocity of the gear is 50 rpm cw, what is the linear acceleration for the gear center and the instant center?

5.25 In Fig. P5.25, determine the linear acceleration for point P. Point P is on a plate which is rigidly fastened to the spur gear. The gear center, at this instant, has an acceleration of 5 ft/sec^2 and a velocity of 10 ft/sec in the same direction. The rack is stationary. The gear diameter is 4 in.

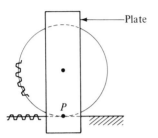

Fig. P5.25.

5.26 In Fig. P5.26, gear G is in mesh with an internal gear and a stationary rack. Determine the angular acceleration for the internal gear. The center of G, point O, is accelerating downward at 5 ft/sec^2.

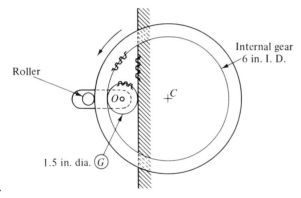

Fig. P5.26.

5.27 Figure P5.27 shows a planetary gear system. At this instant the arm $Ⓐ$ rotates at 1000 rad/sec ccw and accelerates at 500 rad/sec^2 ccw. The ring gear $Ⓡ$ is stationary. The ring gear is 10 in. in diameter, the planet gear $Ⓟ$ is 3 in. in diameter, and $OP = 2.5$ in. Determine the acceleration for contact point X and point P at this instant. Point P is on a triangular piece which is securely fastened to gear $Ⓟ$.

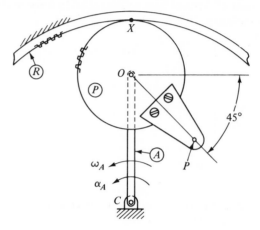

Fig. P5.27.

5.28 Assume that the cylinder shown in Fig. 5.17 has a constant angular velocity of 20 rad/sec ccw and that the piston moves inwardly at 24 in./sec (relative to the cylinder). Distance $XA = 8$ in. Determine the linear acceleration for point Y.

5.29 In Fig. P5.29, at this instant, link $Ⓛ$ is rotating at 1 rad/sec cw with an acceleration of 1 rad/sec^2. For the position shown, determine the linear acceleration for the point follower. The point follower moves horizontally.

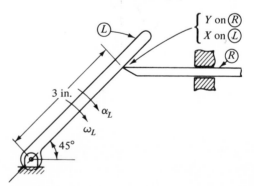

Fig. P5.29.

5.30 A disk ⓡ (Fig. P5.30) rotates at a constant 500 rpm ccw about point B. A pin, fixed to ⓡ, drives the gear sector ⓛ about pivot point A. Sector ⓛ in turn drives the rack ⓜ. For the position shown, determine (a) the angular velocity for the gear sector, (b) the angular acceleration for the gear sector, and (c) the linear acceleration for the rack.

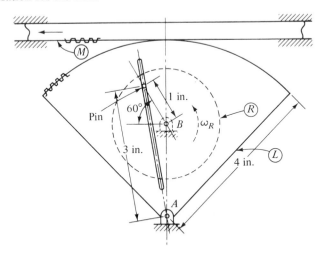

Fig. P5.30.

5.31 For the position shown in Fig. P5.31, determine the angular acceleration for the four-station Geneva. The driver rotates at a constant 60 rpm cw.

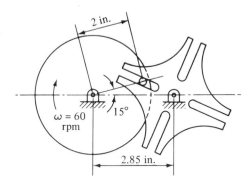

Fig. P5.31.

5.32 In Fig. P5.32, a disk, which pivots about point B, is driven by a pneumatic cylinder. At the instant shown, the piston rod has a constant outward velocity of 15 in./sec relative to the cylinder. Determine the angular velocity and acceleration for the disk.

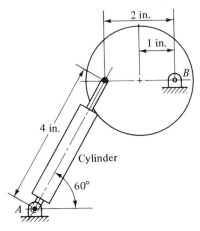

Fig. P5.32.

Chapter **6**

CAMS

6.1 Introduction

A cam is a mechanical device which imparts motion to a follower by means of direct contact. The contour or shape of a cam is dependent on the type of cam and the type of follower. Usually the cam is keyed to a rotating shaft (Fig. 6.1).

6.2 Classification of Cams

Cams are classified according to their physical construction and type of follower. The plate cam, also called the disk cam, is the most popular type of cam. This cam usually resembles a flat plate with the cam shape developed along its periphery. Plate cams are shown in Fig. 6.2 with various types of followers.

Figures 6.2(a) to (d) are grouped as radial or in-line followers. A radial follower moves in translation such that the line of action of the follower passes through the cam axis of rotation. A point follower [Fig. 6.2(a)] is limited to low

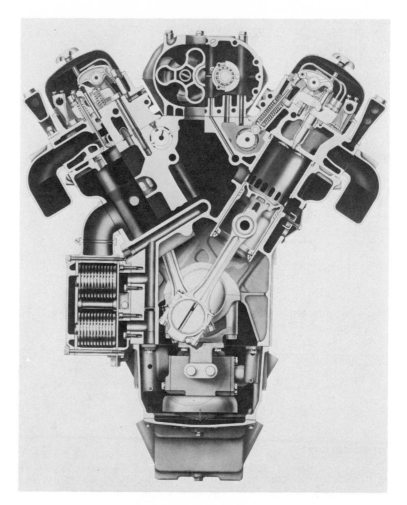

Fig. 6.1. *V-type diesel engine. Cams operate exhaust valves and fuel injectors.* (Courtesy of Detroit Diesel Engine Division of General Motors Corporation.)

speed and low force because of extreme wear. This type of follower is very sensitive to cam shape; thus it is ideal for applications involving abrupt changes in follower motion. Sliding action between the cam and follower is reduced with a roller follower [see Fig. 6.2(b)]. Flat- or spherical-faced followers [Figs. 6.2(c) and (d)] are used for relatively steep cams, as found on automobile valve lifters. Figure 6.2(e) shows an offset follower; the line of action of the follower is offset from the camshaft center. Offsetting of the follower can reduce the steepness (pressure angle) encountered between the cam and follower on the upward stroke. A swinging follower [Fig. 6.2(f)] compared to a translating follower can operate with a larger steepness (pressure angle). The yoke and conjugate cams

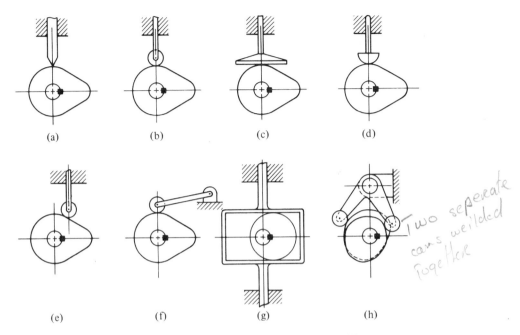

(a)   (b)   (c)   (d)

(e)   (f)   (g)   (h)

Fig. 6.2. Plate cams. (a) *Point or knife-edge follower;* (b) *roller follower;* (c) *flat-face follower;* (d) *circular or spherical-face follower;* (e) *offset roller follower;* (f) *swinging, oscillating, or pivoted follower;* (g) *yoke cam with flat-face followers (also called constant-breadth cam);* (h) *conjugate cam with swinging follower.*

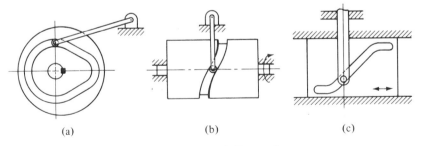

(a)   (b)   (c)

Fig. 6.3. Positive-acting cams. (a) *Face or face-groove cam;* (b) *cylindrical, drum, or barrel cam;* (c) *translating, wedge, or flat-plate cam.*

[Figs. 6.2(g) and (h)] are positive-acting cams which do not need a follower spring.

Other positive-acting cams are shown in Fig. 6.3. A face cam [Fig. 6.3(a)] shows a grooved track on the face of a flat plate. The cylindrical cam [Fig. 6.3(b)] is a popular type of cam. This cam is grooved on a cylindrical surface.

Figure 6.3(c) shows a grooved plate called a translating cam. This cam moves back and forth while driving a follower.

An end cam, shown in Fig. 6.4, has its working surface at the end of a cylinder.

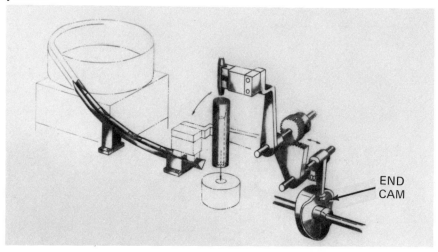

Fig. 6.4. *End cam used in transfer mechanism.* (Courtesy of Matsushita Electric Industrial Company Ltd.)

6.3 Displacement Diagram

This section is concerned with a description of the displacement diagram. How the displacement diagram is developed and specifically how the displacement diagram is used to develop a cam contour will be discussed in later sections. A displacement diagram (Fig. 6.5) is usually a rectangular layout of follower motion related to the cam angle or time. The vertical axis indicates the follower position, and the horizontal axis indicates the angular position of the cam.

Figure 6.5 shows a plate cam with a radial roller follower. The cam is physically marked from 0 to 360 deg in 30-deg increments. The displacement diagram is numbered in the same fashion. Frequently, numbers are used to designate the cam angles; these are also shown on the displacement diagram. To illustrate the relationship between the displacement diagram, cam, and follower, imagine the cam rotated 150 deg ccw from its starting position. The 150-deg mark on the cam would appear under the roller follower, and the roller center would be in position 5. The roller centers are in line with the height of the displacement diagram. Observe that the cam rotates ccw, but that cam angles are sequenced in a cw direction.

The displacement curve is separated into dwell, rise, and fall (or return) portions. During the dwell portion, the follower is at rest. During the rise portion,

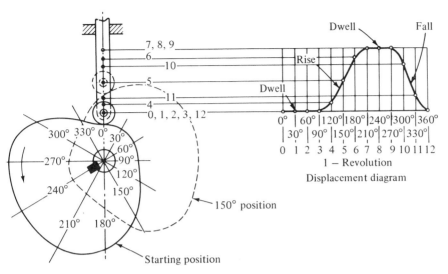

Fig. 6.5. *Plate cam with radial roller follower and displacement diagram.*

the follower is moving upward, and during a fall portion the follower is moving downward. Each 30-deg increment on the displacement diagram is equally spaced because the cam is assumed to rotate at a constant rpm.

6.4 Cam Nomenclature

Before discussing cam terminology, a brief outline on cam construction may be helpful. A cam (Fig. 6.6) is graphically constructed from a base circle. To simplify construction, the base circle is held stationary and rollers are drawn about the base circle at predetermined radial positions. These rollers form an envelope called the cam surface.

Cam terminology follows (refer to Fig. 6.6):

Base circle. The base circle is the smallest circle which can be drawn to the cam surface from the center of rotation. The base circle diameter determines cam size. The base radius is symbolized as R_b.

Trace point. The trace point of a roller follower is the roller center.

Pitch curve. The pitch curve is the path of the trace point. For a radial point follower, the pitch curve coincides with the cam surface.

Cam surface. The cam surface, cam profile, or working curve is the contour of the cam.

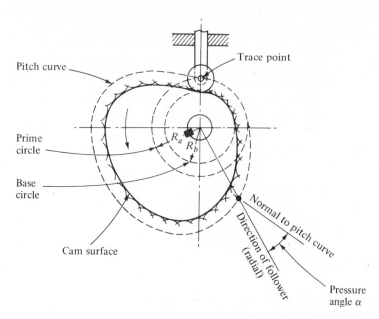

Fig. 6.6. Cam nomenclature.

Prime circle. The prime circle is the smallest circle which can be drawn to the pitch curve from the center of rotation. The prime radius is symbolized as R_a. For Fig. 6.6, $R_a = R_b + R_r$, where R_r is the roller radius.

Pressure angle. The pressure angle is the angle between the normal to the pitch curve and the direction of the follower. For a roller follower, the normal line passes through the roller center and the contact point at the cam surface. A pressure angle that is too large may cause jamming of the follower on the rise portion of the cam. The pressure angle is symbolized as α (alpha).

Pitch point. The pitch point is a point on the pitch curve at which the pressure angle is maximum. The maximum pressure angle is represented as α_m.

6.5 Graphical Cam Layouts

Cams which are manufactured from a graphical layout are inherently inaccurate and therefore are usually used for low-speed application. High-speed cams require a careful choice of follower motion (low change of acceleration) and material and must be very accurately manufactured.

All the cams shown in this section utilize the same displacement diagram, Fig. 6.7. To simplify the cam layout and reduce drafting time, 30-deg increments will be used. If accuracy is required, smaller increments must be used. To lay out a cam, the base circle is held stationary and the follower is constructed about

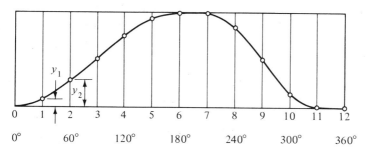

Fig. 6.7. Displacement diagram.

the base circle. This is technically called an inversion. The constructions shown here are not necessarily the only method of construction.

Plate Cam-Radial Point Follower

Known data: Displacement diagram (Fig. 6.7), base circle, and ccw cam rotation. Construction procedure:

STEP 1. Refer to Fig. 6.8. Starting at the follower center line, draw radial lines which divide the base circle into 12 equal parts; use 30-deg increments ($360°/12 = 30°$ increment). This increment corresponds to the divisions shown on the displacement diagram.

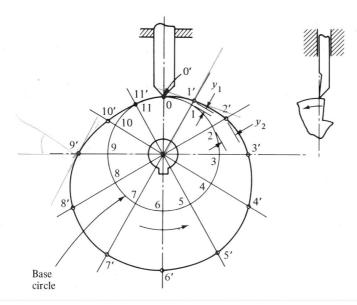

Fig. 6.8. Plate cam-radial point follower.

STEP 2. Label radial lines 0 to 11 in a direction opposite to the cam rotation; start at the follower center line. Note that radial line 12 and 0 are the same.

STEP 3. Using dividers, transfer follower displacement y_1 from the displacement diagram onto radial line 1 at the intersection of the radial line with the base circle. This locates follower point 1′. Transfer displacements until all points are located.

STEP 4. Using a french curve, draw a smooth curve through points 0′, 1′, 2′, to 11′, 0′. This curve is the cam surface.

A point or knife-edge follower is restricted to low rpm and low force because of excessive wear at the follower edge. The point follower is very sensitive to cam contour. Figure 6.8 shows a modified translating point follower which allows the follower to fall abruptly.

Plate Cam-Radial Roller Follower

Known data: Displacement diagram (Fig. 6.7), base circle, roller diameter, and ccw cam rotation. Construction procedure:

STEP 1. Refer to Fig. 6.9. Draw the prime circle. Draw 12 equally spaced radial lines. Label the lines 0 to 11 in cw direction.

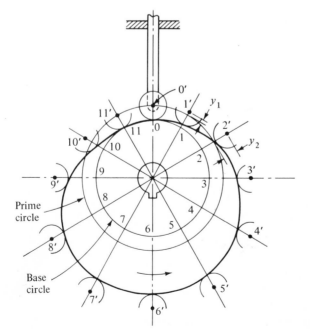

Fig. 6.9. Plate cam-radial roller follower.

STEP 2. Transfer follower displacement y_1 from the displacement diagram onto radial line 1 at the prime circle. This locates roller center 1'. Transfer displacements until all roller centers are located.

STEP 3. Draw a roller circle from each roller center.

STEP 4. Draw a smooth curve tangent to each roller circle. This curve is the cam surface.

A roller follower is used to reduce friction between the cam and follower. Needle and ball bearings are usually used for roller followers.

Plate Cam-Radial Flat-Face Follower

Known data: Displacement diagram (Fig. 6.7), base circle, and cw cam rotation. Construction procedure:

STEP 1. Refer to Fig. 6.10. Draw 12 equally spaced radial lines. Label the lines 0 to 11 in ccw direction.

STEP 2. Transfer follower displacement y_1 from the displacement diagram onto radial line 1 at the base circle. This locates the center point on the follower face; label it point 1'. Transfer displacements until all center points are located.

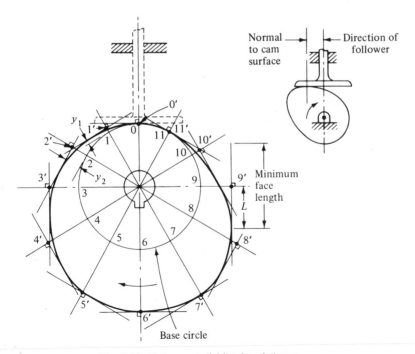

Fig. 6.10. Plate cam-radial flat-face follower.

STEP 3. Draw a line perpendicular to radial line 1 which passes through point 1'. Construct perpendicular lines at other positions.

STEP 4. Draw a smooth curve tangent to each perpendicular line. This curve is the cam surface.

The cam layout is not complete until the face length of the follower is determined. Using dividers, examine each perpendicular line to ascertain the longest length from the follower center point to the contact point with the cam. This scanning shows that the longest length is at position 9, length marked L. The minimum face length of a symmetrical follower must be at least twice L. To provide for some margin of safety, use a dimension larger than the minimum face length. It should be noted that in Fig. 6.10 the pressure angle for a flat-face follower is zero. That is, the angle between the normal to the cam surface and the direction of the follower is zero. This considerably reduces the possibility of jamming. For small cams, a flat-face follower is sometimes preferred over a radial roller follower because of excessive stress on the roller pin.

Plate Cam-Offset Roller Follower

Known data: Displacement diagram (Fig. 6.7), base circle, eccentricity, roller diameter, and ccw cam rotation. Construction procedure:

STEP 1. Refer to Fig. 6.11. Draw a circle (offset circle) which is tangent to the follower center line. Starting at the tangent point, draw short radial lines which divide the offset circle into 12 equal parts. Label the lines 0 to 11 in cw direction.

STEP 2. Draw a line perpendicular to radial line 1 which is tangent to the offset circle. This perpendicular line represents the line of action of the follower; label it line 1 . Construct perpendicular lines at other positions.

STEP 3. Transfer follower displacement y_1 from the displacement diagram onto the follower center line 0' at the center of the roller follower. This locates roller center 1''. Transfer displacements until all roller centers are located on the follower center line.

STEP 4. Using the cam center as the center of rotation, draw an arc through point 1'' which intersects line 1'. This locates roller center 1'''. Locate all other roller centers about the base circle.

STEP 5. From each roller center, draw a roller circle. Draw a smooth cam curve tangent to each roller circle.

Generally speaking, the follower stem of a translating follower tends to jam in the follower guide on the rise portion because of a sideward push on the follower from the rotating cam. Jamming can be reduced by decreasing the pressure angle. Offsetting usually decreases the pressure angle. To decrease the pressure angle on the rise portion, offset the follower to the left of the cam center

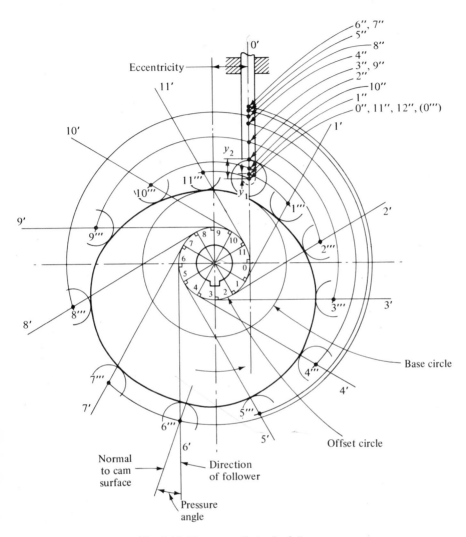

Fig. 6.11. *Plate cam-offset roller follower.*

for a cw cam rotation, or offset the follower to the right for a ccw cam rotation; refer to Sect. 6.12. Note that this offsetting increases the pressure angle on the fall (or return) portion; however, this is usually of little consequence.

Plate Cam-Swinging Roller Follower

Known data: Displacement diagram (Fig. 6.7), base circle, roller diameter, follower arm length, follower pivot location P_0, and ccw cam rotation. It is assumed that the displacement diagram is a plot of the follower position on the arc of motion vs. the cam angle. Construction procedure:

STEP 1. Refer to Fig. 6.12. Draw radial line 0 which passes through the roller center. Starting from radial line 0, draw radial lines which divide the base circle into 12 equal parts. Label the lines 1 to 11 in cw direction.

STEP 2. Draw the arc of motion for the center of the roller follower. Transfer follower displacement y_1 from the displacement diagram onto the arc of

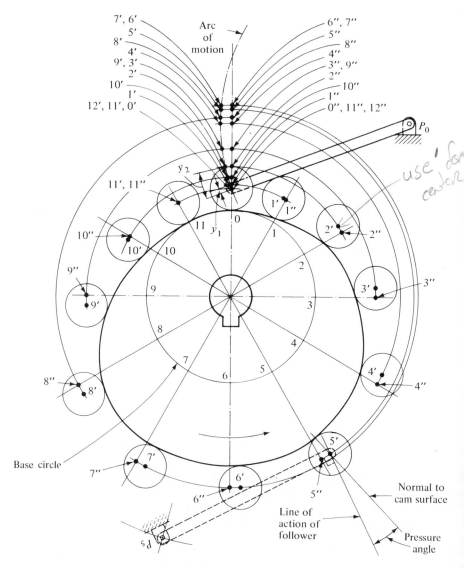

Fig. 6.12. Plate cam-swinging roller follower.

motion at the center of roller $0'$. This locates roller center $1'$. Transfer displacements until all roller centers are located on the arc of motion. Note that to minimize error, split large displacements into several small chordal lengths.

STEP 3. Using the cam center as the center of rotation, draw an arc through point $1'$ which intersects radial lines 0 and 1. Label both intersection points $1''$. Locate all other intersection points, such as point $2''$ on radial line 0 and point $2''$ on radial line 2, etc.

STEP 4. At position 0, transfer small chordal offset $1'1''$ onto the arc at point $1''$, where point $1''$ is on radial line 1. This locates roller center $1'$ at position 1. Locate all other roller centers about the base circle. Note that each chordal offset is left of its respective radial line (when viewed from the center of the cam).

STEP 5. From each roller center, draw a roller circle. Draw a smooth cam curve tangent to each roller circle.

A swinging follower can tolerate a large pressure angle because the sideward push on the follower (push is essentially along the normal line) does not tend to jam the follower. To avoid locking, the normal line should not pass through or near the follower pivot point. Figure 6.12 shows the pressure angle at position 5. To locate the follower pivot P_5, move point P_0 150 deg cw. The line of action of the follower is drawn perpendicular to the follower arm.

Plate Cam-Swinging Flat-Face Follower

Known data: Displacement diagram (Fig. 6.7), base circle, follower dimensions, follower pivot location P_0, and ccw cam rotation. It is assumed that the displacement diagram is a plot of the follower position on the arc of motion vs. the cam angle. The displacement diagram is coordinated to point $0'$ on the follower surface. Construction procedure:

STEP 1. Refer to Fig. 6.13. Draw radial line 0 through follower pivot point P_0. Starting from radial line 0, draw radial lines which divide the base circle into 12 equal parts. Label the lines 1 to 11 in cw direction.

STEP 2. Using the cam center as the center of rotation, draw a circle through P_0. The intersection of this circle with the radial lines locates follower pivot points P_1, P_2, P_3, etc.

STEP 3. At P_0, draw a circle which is tangent to the face of the follower. Call this circle the tangent circle. Draw this circle at the other follower pivot points.

STEP 4. Using P_0 as the center, draw the arc of motion through point $0'$. Transfer follower displacement y_1 from the displacement diagram onto the arc of motion at point $0'$. This locates a point on the follower face; label it point $1'$. Transfer displacements until all follower face points are located.

STEP 5. Using radius R $(R = \overline{P_0 0'})$, draw an arc from point P_1. Now, using the cam center as the center of rotation, draw an arc through point $1'$ which

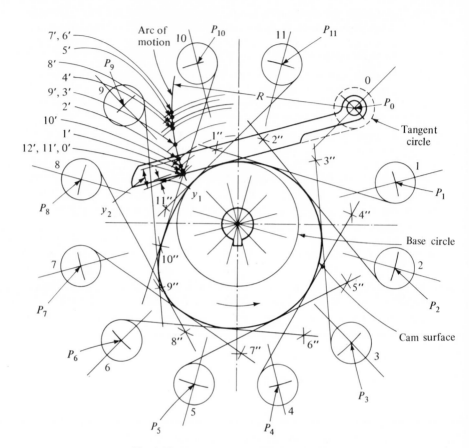

Fig. 6.13. *Plate cam-swinging flat-face follower.*

intersects the arc previously drawn. The intersection locates a point on the follower face; label it point $1''$. Locate all other follower face points.

STEP 6. Draw a line which passes through point $1''$ and is tangent to the tangent circle at location P_1. This line represents the follower face. In a similar fashion, construct the other positions of the follower face.

STEP 7. Draw a smooth cam curve tangent to each follower face. A check of the face length, particularly at position 10, shows the assumed length to be adequate.

EXAMPLE 6.1

Given: Figure 6.14(a); swinging roller follower link ABC oscillates through 18 deg. Each position of BC corresponds to a cam angular displacement of 30 deg. $AB = 2$ in., $BC = 2.5$ in., the roller diameter $= .5$ in., the base circle diameter $= 2$ in., and the cam rotates ccw.

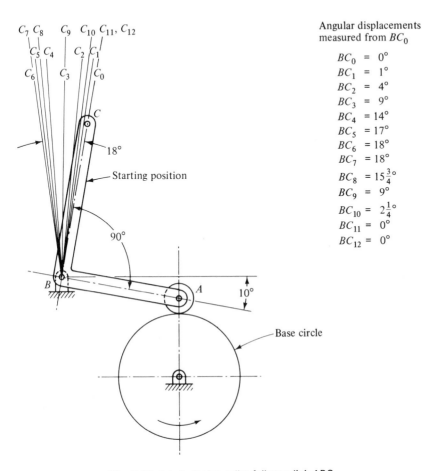

Angular displacements
measured from BC_0

$BC_0 = 0°$
$BC_1 = 1°$
$BC_2 = 4°$
$BC_3 = 9°$
$BC_4 = 14°$
$BC_5 = 17°$
$BC_6 = 18°$
$BC_7 = 18°$
$BC_8 = 15\frac{3}{4}°$
$BC_9 = 9°$
$BC_{10} = 2\frac{1}{4}°$
$BC_{11} = 0°$
$BC_{12} = 0°$

Fig. 6.14. (a) *Swinging roller follower link* ABC.

Determine: Cam contour.

Solution. The construction is shown only for three positions.

STEP 1. Refer to Fig. 6.14(b). Draw radial line 0 which passes through the roller center. Starting from radial line 0, draw radial lines which divide the base circle into 12 equal parts. Label the lines 1 to 11 in cw direction.

STEP 2. Draw the arc of motion for the center of the roller follower which intersects the positions of BC. Label the intersection points C_0', C_1', C_2', etc. Transfer these points onto the arc of motion at the center of roller 0'. Label the roller centers 0', 1', 2', etc.

STEP 3. Continue the construction starting from step 3 for the plate cam-swinging roller follower (refer to page 195).

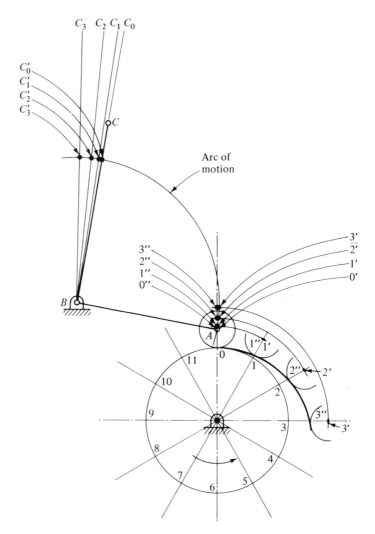

Fig. 6.14. (Cont.) (b) Cam contour.

6.6 Positive-Acting Cams

A positive-acting cam does not require a follower spring because the cam itself constantly pushes the follower. Excessive clearance, which results from wear and/or manufacturing inaccuracies, between the cam and follower can cause impact. This shock action is especially detrimental at high speeds.

A yoke cam is a single plate cam enclosed by two followers. Followers may be translating or swinging type and either flat-face or roller. The follower surfaces or roller centers are a constant distance apart. Figure 6.15 shows a

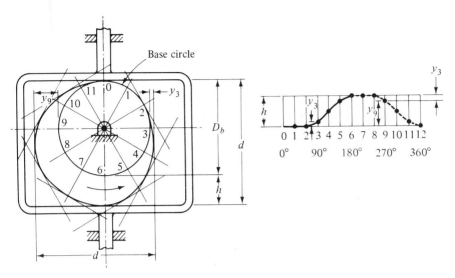

Fig. 6.15. Yoke cam with translating flat-face followers.

layout of a yoke cam with translating flat-face followers. Distance d between follower faces is equal to the base circle diameter D_b plus the maximum follower displacement h. Construction of a yoke cam requires only half a displacement diagram (0 to 180 deg). Position 3 along with position 9 will be used to illustrate the construction procedure. Lay out y_3 from the displacement diagram onto the base circle. Construct the follower face perpendicular to radial line 3. The opposite face (at position 9) is located a distance d from this face.

6.7 Follower Motion

At very low speeds almost any type of follower motion is suitable. For medium-speed cams, three types of motion frequently used are parabolic motion, simple harmonic motion (SHM), and cycloidal motion. Of the three motions, cycloidal motion gives the best overall performance, providing the cam is manufactured accurately. At high speed, cycloidal, modified trapezoidal, and modified sinusoidal, to name a few, are employed. This text will be limited to a discussion of SHM and cycloidal follower motion. For other motions, the reader is referred to *Cams* by H. A. Rothbart (John Wiley & Sons, Inc., New York, 1956) and *Cam Design and Manufacture* by P. W. Jensen (Industrial Press, New York, 1965).

Follower displacements for SHM and cycloidal motion can be determined graphically, mathematically, or from a table. The graphical technique lacks accuracy, but it is fast. The mathematical method, discussed in this section, is time consuming unless a computer is used. And the table method, discussed in Sect. 6.10 is accurate but is limited to 120 points for the rise or the fall portion.

Figure 6.16(a) shows the rise portion of a displacement diagram. This particular curve happens to be cycloidal and can be described mathematically as

$$y = \frac{h\theta}{\beta} - \frac{h}{2\pi} \sin \frac{2\pi\theta}{\beta}$$

(6.1)

where y = displacement of the follower, in.
h = maximum displacement of the follower, in.
θ = cam angle corresponding to follower displacement y, rad
β = cam angle corresponding to total rise h, rad

To illustrate how the equation is used, let $\theta = 60$ deg, $\beta = 180$ deg, and
$h = 1.50$ in. Determine y; refer to Fig. 6.16(a). Substituting into Eq. (6.1),

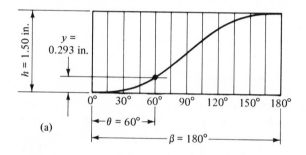

(a)

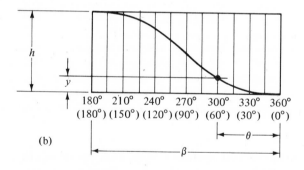

(b)

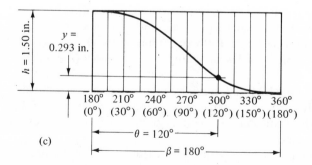

(c)

Fig. 6.16. Displacement diagram or follower motion for
(a) rise, (b) fall, (c) fall.

$$y = \frac{1.5(60° \times \pi/180°)}{180° \times \pi/180°} - \frac{1.5}{2\pi} \sin \frac{2\pi(60° \times \pi/180°)}{(180° \times \pi/180°)}$$

$$y = 1.5 \left(\frac{6}{18}\right) - \frac{1.5}{2\pi} \sin \frac{2\pi}{3} \tag{1}$$

To evaluate the sin function, it is advisable to change $2\pi/3$ rad into degrees. Note that when π appears outside the trigonometric function it is not converted into 180 deg. Now Eq. (1) appears as

$$y = 1.5 \left(\frac{6}{18}\right) - \frac{1.5}{2\pi} \sin \frac{2(180°)}{3}$$

and

$$y = .500 - .207 = .293 \text{ in.}$$

The rise equation can be used for fall motion [Fig. 6.16(b)] providing the cam angles are renumbered starting with 0 deg from the smaller end. Observe that in Fig. 6.16(b) the angle θ is referenced from the right end.

Another way of calculating the follower displacement for the fall portion is to use the fall equation. The fall equation for cycloidal motion is

$$y = h - \frac{h\theta}{\beta} + \frac{h}{2\pi} \sin \frac{2\pi\theta}{\beta} \tag{6.2}$$

To use this equation properly, the cam angles must be renumbered starting with 0 deg from the larger end.

Let us solve an example using Eq. (6.2). Determine y [cycloidal motion, Fig. 6.16(c)] for a cam angle of 300 deg, $\beta = 180$ deg, and $h = 1.50$ in. The 300 deg is referenced from the beginning of the full displacement diagram. First, renumber the angles in Fig. 6.16(c): Change 180 deg to 0 deg, 210 deg to 30 deg, etc. Cam angle 300 deg corresponds to $\theta = 120$ deg. Now, substitute into Eq. (6.2):

$$y = 1.500 - \frac{1.5(120° \times \pi/180°)}{180° \times \pi/180°} + \frac{1.5}{2\pi} \sin \frac{2\pi(120° \times \pi/180°)}{(180° \times \pi/180°)}$$

$$y = 1.500 - 1.000 - .207 = .293 \text{ in.}$$

Note that this is exactly the same answer we obtained in the previous example for the rise portion.

6.8 Simple Harmonic Motion

An example of SHM is the motion of point P in Fig. 6.17, where P is the horizontal projection of point A. Here link OA rotates at a constant angular

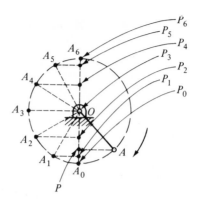

Fig. 6.17. Rotating link OA. Motion of P is SHM.

velocity about O. To help visualize the motion of P, several equally spaced positions of OA are shown: OA_0, OA_1, OA_2, etc. Figure 6.17 forms the basis of a graphical layout for SHM.

Figure 6.18 shows a graphical layout of a SHM displacement diagram (rise portion only). To outline the construction, assume that $h = 1$ in. and $\beta = 90$ deg.

STEP 1. Choose a reasonable cam angle which divides evenly into β. A 10-deg increment will be used here. Now, divide the θ axis into nine equal parts ($\beta/10° = 90°/10° = 9$).

STEP 2. Draw a 1-in. diameter harmonic semicircle. Divide the semicircle into nine equal parts. Use dividers or a harmonic angle of 20 deg ($180°/9 = 20°$).

STEP 3. Project the points horizontally from the semicircle to the corresponding cam angles. Draw in the complete displacement curve.

The fall portion of a SHM a displacement diagram is handled in a similar fashion.

The displacement diagram can be plotted from calculated data obtained from

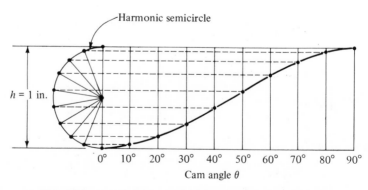

Fig. 6.18. Graphical layout of SHM displacement diagram.

the SHM equation. The SHM displacement equation, rise portion, is

$$y = \frac{h}{2}\left(1 - \cos\frac{\pi\theta}{\beta}\right) \tag{6.3}$$

The fall equation is

$$y = \frac{h}{2}\left(1 + \cos\frac{\pi\theta}{\beta}\right) \tag{6.4}$$

Refer to Sect. 6.7 for the meaning of the symbols.

6.9 Cycloidal Motion

Figure 6.19 shows a circle rolling at constant angular velocity, without slip, on a vertical line. Point A is a point on the circumference, and point P is the horizontal projection of A. Point A traces out a cycloid, and the motion of point P, such as, P_0, P_1, P_2, etc., is called cycloidal motion. The positions of P and A are spaced at equal time intervals.

Cycloidal motion starts and ends smoothly because of low accelerations at the start and end of motion.

The accuracy required for cycloidal motion practically eliminates a graphical

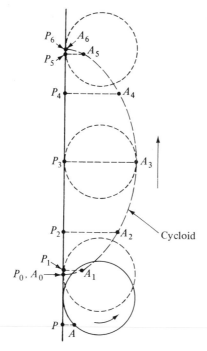

Fig. 6.19. Circle rolling on line. Motion of P is cycloidal.

approach; therefore the graphical layout for the displacement diagram will not be shown. Cycloidal displacements can be obtained from the displacement table (Sect. 6.10) or from the cycloidal equation.

The cycloidal displacement equation, rise portion, is

$$y = \frac{h\theta}{\beta} - \frac{h}{2\pi} \sin \frac{2\pi\theta}{\beta} \tag{6.5}$$

The fall equation is

$$y = h - \frac{h\theta}{\beta} + \frac{h}{2\pi} \sin \frac{2\pi\theta}{\beta} \tag{6.6}$$

Refer to Sect. 6.7 for the meaning of the symbols.

6.10 Displacement Table for Follower Motion

Follower displacement can be obtained accurately from the displacement table (Fig. 6.20). This table is limited to 120 points for the rise or fall portion. Cam angle θ is obtained from $\theta = (\beta/120)(\phi)$, and the follower displacement is $y = (h/1)(f)$, where $\beta/120$ can be called the angular factor and $h/1$ or h the displacement factor. The dimensions of θ and y are determined by the dimensions of β and h; ϕ and f are dimensionless. For example, if β is in degrees, then θ is in degrees.

Let us solve a problem using the table. Determine the follower displacements (rise only) for cycloidal motion; $\beta = 90$ deg, and $h = 1.5$ in. The angular factor is

$$\frac{\beta}{120} = \frac{90°}{120} = \frac{3°}{4}$$

To obtain cam angle θ, multiply the angular factor by ϕ:

$$\theta_0 = 0(\tfrac{3}{4}) = 0°$$
$$\theta_1 = 1(\tfrac{3}{4}) = .75°$$
$$\theta_2 = 2(\tfrac{3}{4}) = 1.5°$$
$$\text{etc.}$$

The displacement factor is 1.5 in. Follower displacement y is obtained by multiplying the displacement factor by f:

$$y_0 = .000000(1.5) = .000000 \text{ in.}$$
$$y_1 = .000003(1.5) = .000005 \text{ in.}$$
$$y_2 = .000030(1.5) = .000045 \text{ in.}$$
$$\text{etc.}$$

$$\theta = \frac{\beta}{120°} \, (\phi) \qquad \text{and} \qquad y = \frac{h}{1} \, (f)$$

Simple Harmonic

ϕ	f	ϕ	f	ϕ	f
1	0.000171	41	0.261420	81	0.761249
2	0.000685	42	0.273004	82	0.772319
3	0.001541	43	0.284744	83	0.783203
4	0.002739	44	0.296631	84	0.793892
5	0.004277	45	0.308658	85	0.804380
6	0.006155	46	0.320816	86	0.814660
7	0.008372	47	0.333096	87	0.824724
8	0.010926	48	0.345491	88	0.834565
9	0.013815	49	0.357992	89	0.844177
10	0.017037	50	0.370590	90	0.853553
11	0.020590	51	0.383277	91	0.862687
12	0.024471	52	0.396044	92	0.871572
13	0.028679	53	0.408882	93	0.880202
14	0.033209	54	0.421782	94	0.888572
15	0.038060	55	0.434736	95	0.896676
16	0.043227	56	0.447735	96	0.904508
17	0.048707	57	0.460770	97	0.912063
18	0.054496	58	0.473832	98	0.919335
19	0.060591	59	0.486911	99	0.926320
20	0.066987	60	0.500000	100	0.933012
21	0.073679	61	0.513088	101	0.939408
22	0.080664	62	0.526167	102	0.945503
23	0.087936	63	0.539229	103	0.951292
24	0.095491	64	0.552264	104	0.956772
25	0.103323	65	0.565263	105	0.961939
26	0.111427	66	0.578217	106	0.966790
27	0.119797	67	0.591117	107	0.971320
28	0.128427	68	0.603955	108	0.975528
29	0.137312	69	0.616722	109	0.979409
30	0.146446	70	0.629409	110	0.982962
31	0.155822	71	0.642007	111	0.986184
32	0.165434	72	0.654508	112	0.989073
33	0.175275	73	0.666903	113	0.991627
34	0.185339	74	0.679183	114	0.993844
35	0.195619	75	0.691341	115	0.995722
36	0.206107	76	0.703368	116	0.997260
37	0.216796	77	0.715255	117	0.998458
38	0.227680	78	0.726995	118	0.999314
39	0.238750	79	0.738579	119	0.999828
40	0.250000	80	0.750000	120	1.000000

Cycloidal

ϕ	f	ϕ	f	ϕ	f
1	0.000003	41	0.208188	81	0.816808
2	0.000030	42	0.221240	82	0.828728
3	0.000102	43	0.234646	83	0.840250
4	0.000243	44	0.248391	84	0.851365
5	0.000474	45	0.262460	85	0.862065
6	0.000818	46	0.276837	86	0.872343
7	0.001297	47	0.291507	87	0.882195
8	0.001932	48	0.306451	88	0.891616
9	0.002745	49	0.321651	89	0.900603
10	0.003755	50	0.337089	90	0.909154
11	0.004984	51	0.352745	91	0.917270
12	0.006451	52	0.368599	92	0.924949
13	0.008173	53	0.384630	93	0.932195
14	0.010171	54	0.400818	94	0.939010
15	0.012460	55	0.417141	95	0.945398
16	0.015058	56	0.433576	96	0.951365
17	0.017980	57	0.450102	97	0.956917
18	0.021240	58	0.466697	98	0.962061
19	0.024854	59	0.483337	99	0.966808
20	0.028834	60	0.500000	100	0.971165
21	0.033191	61	0.516662	101	0.975145
22	0.037938	62	0.533302	102	0.978759
23	0.043082	63	0.549897	103	0.982019
24	0.048634	64	0.566423	104	0.984941
25	0.054601	65	0.582859	105	0.987539
26	0.060989	66	0.599181	106	0.989828
27	0.067804	67	0.615369	107	0.991826
28	0.075050	68	0.631400	108	0.993548
29	0.082729	69	0.647254	109	0.995015
30	0.090845	70	0.662910	110	0.996244
31	0.099396	71	0.678348	111	0.997254
32	0.108383	72	0.693548	112	0.998067
33	0.117804	73	0.708492	113	0.998702
34	0.127656	74	0.723162	114	0.999181
35	0.137934	75	0.737539	115	0.999525
36	0.148634	76	0.751608	116	0.999756
37	0.159749	77	0.765353	117	0.999897
38	0.171271	78	0.778759	118	0.999969
39	0.183191	79	0.791811	119	0.999996
40	0.195501	80	0.804498	120	1.000000

Fig. 6.20. *Displacement table for simple harmonic and cycloidal motion.*

205

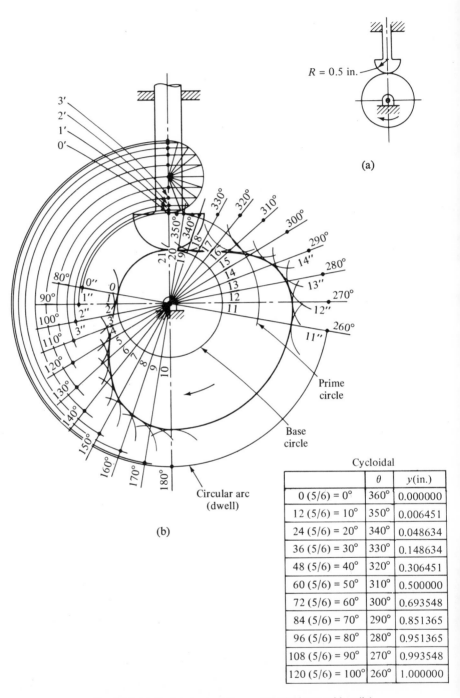

$R = 0.5$ in.

(a)

Prime circle

Base circle

Circular arc (dwell)

(b)

Cycloidal

	θ	y(in.)
0 (5/6) = 0°	360°	0.000000
12 (5/6) = 10°	350°	0.006451
24 (5/6) = 20°	340°	0.048634
36 (5/6) = 30°	330°	0.148634
48 (5/6) = 40°	320°	0.306451
60 (5/6) = 50°	310°	0.500000
72 (5/6) = 60°	300°	0.693548
84 (5/6) = 70°	290°	0.851365
96 (5/6) = 80°	280°	0.951365
108 (5/6) = 90°	270°	0.993548
120 (5/6) = 100°	260°	1.000000

Fig. 6.21. (a) *Radial circular follower.* (b) *Plate cam with radial circular follower and cycloidal displacements.*

206

EXAMPLE 6.2

Given: A radial circular follower [Fig. 6.21(a)] moves through a total displacement of 1 in. Follower dwells for 80 deg of cam rotation, rises with SHM in 100 deg, dwells for 80 deg, and falls with cycloidal motion in 100 deg. The follower radius $= .5$ in., the base circle diameter $= 1.5$ in., and the cam rotates cw.

Determine: Plate cam layout.

Solution

STEP 1. Refer to Fig. 6.21(b). Graphically establish the follower displacements for SHM. Use a cam angle increment of 10 deg. Thus, ten divisions are required $(100°/10° = 10)$.

STEP 2. Draw a 1-in. diameter harmonic semicircle onto the follower centerline. Divide the semicircle into ten equal parts; use a harmonic angle of 18 deg $(180°/10 = 18°)$.

STEP 3. Project points horizontally from the semicircle onto the follower centerline. This locates the centers of the circular follower. Label the centers $0', 1', 2', \ldots, 9', 10'$.

STEP 4. Using the displacement table (Fig. 6.20), determine the follower displacements for cycloidal motion. The angular factor is $100°/120 = 5°/6$, and the displacement factor is 1 in. Follower displacements are calculated and listed in Fig. 6.21(b).

STEP 5. Label radial lines every 10 deg from 80 deg to 180 deg and from 260 deg to 360 deg.

STEP 6. Using the cam center as the center of rotation, draw an arc through point $0'$ which intersects the 80-deg radial line. This locates circular center $0''$. In a similar fashion, locate centers $1''$ to $10''$.

STEP 7. Draw the prime circle from 260 deg to 360 deg. Using data listed in Fig. 6.21(b); on the 260-deg radial line, measure off 1.00 in. from the prime circle. This locates circular center $11''$. Locate other centers $12''$ to $21''$ in a similar fashion.

STEP 8. Draw a circular arc from each circular center. To complete the cam, draw a smooth cam curve tangent to each circular arc. Note that the cam contours from 0 deg to 80 deg and from 180 deg to 260 deg are circular arcs. It is apparent that the cycloidal accuracy cannot be fully utilized in a graphical layout.

6.11 Pressure Angle and Cam Size

Refer to Fig. 6.22. Ideally the push or force F exerted onto the follower by the cam acts along normal line NN. Force F pushes the follower vertically upward and at the same time tends to twist the follower stem about the guide. This twisting action can cause jamming or excessive wear on the follower stem and guide. One means of reducing the magnitude of this twisting action is to decrease the pressure angle. The pressure angle can be decreased by the following methods:

1. Increase the base circle diameter, D_b. This increases cam size.
2. Increase the cam angle, β, for the rise portion.

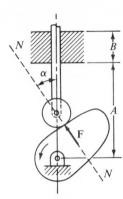

Fig. 6.22. Force F exerted onto the follower by the cam.

3. Decrease the maximum follower displacement, h.
4. Change the amount of offset; refer to Sect. 6.12.
5. Change the type of follower motion. For example, if the choice were between a radial follower moving with SHM or cycloidal motion, a smaller pressure angle is obtained with SHM.
6. Use a secondary follower, as in Fig. 6.23.

Although relatively large pressure angles are being used successfully, it is generally accepted to limit a translating follower to a pressure angle of 30 deg and a swinging follower to 45 deg on the rise portion. Jamming does not occur during the fall portion; hence the pressure angle is not restricted to a 30 or 45 deg maximum.

Additional steps can be taken to improve cam performance:

1. Minimize friction and clearance between moving parts wherever possible.
2. Distance A (Fig. 6.22) should be made as small as possible, and bearing length B as large as possible.
3. The follower stem should be made as rigid as possible.

Generally a small cam is desirable because it requires less machine space. One method of determining cam size is by a trial-and-error procedure. Starting from a cam layout, determine (by measuring) the maximum pressure angle on

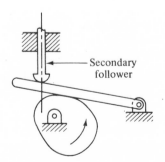

Fig. 6.23. Secondary follower.

the rise portion. If the pressure angle is too large, the common practice is to increase the cam size (base circle) and then lay out another cam. If the pressure angle is too small, the cam can be decreased in size. Assuming another cam is laid out, again determine the maximum pressure angle and, again, if necessary, adjust the cam size. Continue this procedure until a satisfactory pressure angle is obtained. A faster method of determining cam size for translating follower types is discussed in Sect. 6.12.

6.12 Cam Size Determined by Pressure Angle Nomograms

The nomograms shown in Figs. 6.24 and 6.25 are used to determine cam size, or, more specifically, the prime radius R_a for a cam which operates with a translating follower. Also, for a given cam and translating follower, the nomograms can be used to determine the maximum pressure angle for the rise and fall motion. Cams operating with flat-face followers are not included in this discussion because their pressure angles are zero.

The prime radius will be indicated on the nomograms as distance O_1R. Point O_1 represents the cam center or center of rotation and point R the center of the follower. If the follower is a roller type, R is the center of the roller. For a point follower, R represents the tip of the follower. The vertical axis through R is the line of action of the follower. For a radial-type follower, O_1 and R are vertically in line. For an offset-type follower, O_1 is either left or right of the line of action of the follower. The amount of offset (eccentricity) is indicated on the horizontal line labeled e. A unit length on the nomogram is equivalent to the maximum follower displacement h; thus the scale factor is $K = h/1$ unit.

As an example, determine the prime radius R_a for a cam which operates with a radial roller follower where $h = 2.5$ in., β (rise) $= 100$ deg, α_m (rise) $= 30$ deg, and the motion is cycloidal. The direction of cam rotation is not required for this problem.

STEP 1. Use the left side of the cycloidal nomogram (Fig. 6.25). Draw a line tangent to the $\beta = 100$ deg curve which makes an angle of $\alpha = 30$ deg with the vertical.

STEP 2. The intersection of this line with the vertical axis locates cam center O_1. The prime radius is distance O_1R, where O_1R is 1.48 units. Thus, $R_a = O_1R(K) = 1.48$ units (2.5 in./1 unit) $= 3.70$ in.

As another example, determine the prime radius for a cam which operates with a translating roller follower where $h = 2.5$ in., β (rise) $= 100$ deg, α_m (rise) $= 30$ deg, β (fall) $= 80$ deg, α_m (fall) $= 50$ deg, both rise and fall motion is cycloidal, and the cam rotates ccw. Note that the rise portion information is the same as in the previous example.

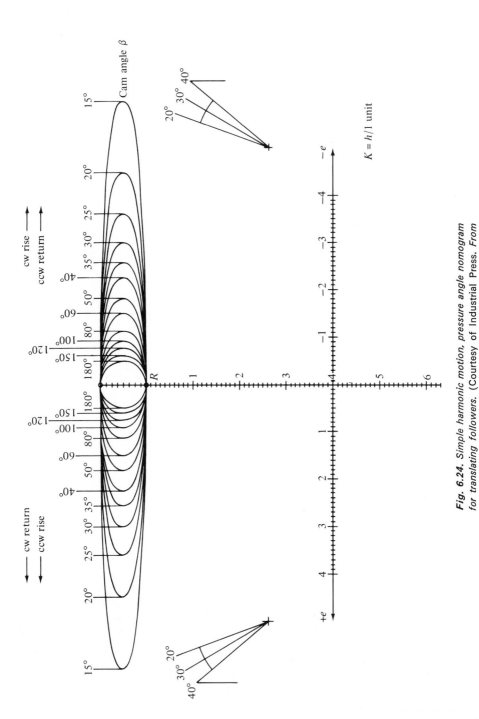

Fig. 6.24. Simple harmonic motion, pressure angle nomogram for translating followers. (Courtesy of Industrial Press. From Cam Design and Manufacture by P. W. Jensen.)

$K = h/1$ unit

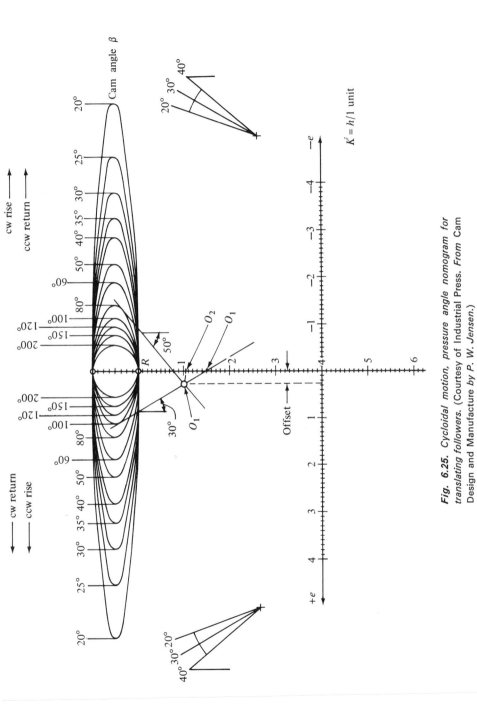

Fig. 6.25. *Cycloidal motion, pressure angle nomogram for translating followers.* (Courtesy of Industrial Press. *From* Cam Design and Manufacture *by P. W. Jensen*.)

STEP 1. For the rise portion, use the left side of the cycloidal nomogram (Fig. 6.25) because the cam is rotating ccw. Draw a line tangent to the $\beta = 100$ deg curve which makes an angle of $\alpha = 30$ deg with the vertical.

STEP 2. For the fall portion, use the right side of the nomogram. Draw a line tangent to the $\beta = 80$ deg curve at an angle of $\alpha = 50$ deg with the vertical.

STEP 3. The intersection of these lines locates cam center O_1. Do not confuse this O_1 with the O_1 on the vertical axis. To determine the scale distance for $O_1 R$; use point R as the center of rotation, swing an arc through O_1 which intersects the vertical axis. Label the intersection point O_2. $O_2 R = 1.06$ units. $R_a = O_2 R(K) = 1.06(2.5) = 2.65$ in.

STEP 4. Determine the offset. The follower offset appears to the right of the cam center, $e = .28(2.5) = .70$ in.

In each of the foregoing examples the cam layout cannot be made until a roller diameter is decided upon.

Comparison of the results obtained from these two examples shows that offsetting the follower reduced the cam size. Note that both cams reach the same maximum pressure angle on the rise. Offsetting the follower in the proper direction can be used to reduce the cam size or to reduce the maximum pressure angle on the rise portion, or both. Generally for cw cam rotation, one offsets the follower to the left of the cam center. For ccw cam rotation, one offsets the follower to the right of the cam center.

It must be emphasized that these nomograms are to be used only for plate cams operating with translating followers, either radial or offset type.

EXAMPLE 6.3

Given: A translating roller follower rises with SHM in 120 deg of cam rotation, dwells in 60 deg, then falls with SHM in 40 deg, and finally dwells for 140 deg. The maximum follower displacement $= 1.5$ in., the follower diameter $= 1$ in., and the cam rotates cw. The pressure angle is limited to 30 deg for rise and 50 deg for fall.

Determine: Displacement diagram and plate cam layout.

Solution

STEP 1. Refer to the nomogram sketch [Fig. 6.26(b)]. For the rise portion, use the right side of the SHM nomogram (Fig. 6.24). Draw a line tangent to the $\beta = 120$ deg curve which makes an angle of $\alpha = 30$ deg with the vertical.

STEP 2. For the fall portion, use the left side of the nomogram. Draw a line tangent to the $\beta = 40$ deg curve at an angle of $\alpha = 50$ deg with the vertical.

STEP 3. The intersection of these lines locates cam center O_1. Using R as the center, swing an arc through O_1 which intersects the vertical axis at O_2. $R_a = O_2 R(K) = 1.35(1.5) = 2.03$ in. The follower offset is to the right of the cam center, $e = .22(1.5) = .33$ in.

STEP 4. Refer to the displacement diagram [Fig. 6.26(a)]. For the rise portion, use a 15-deg increment; mark off eight equal divisions on the θ axis ($120°/15° = 8$). Draw a

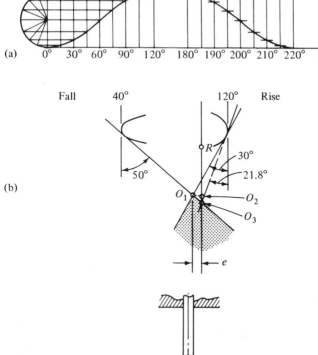

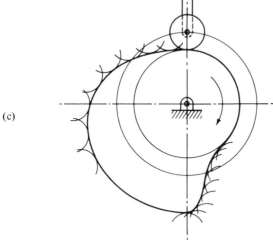

Fig. 6.26. Displacement diagram, nomogram, and cam layout.

1.5-in. diameter harmonic semicircle. Divide the semicircle into eight equal parts. Project points horizontally to the corresponding cam angles.

STEP 5. Skip space on the displacement diagram for the 60-deg dwell. For the fall portion, use a 5-deg increment and eight equal divisions on the θ axis ($40°/5° = 8$).

The fall portion utilizes the same points on the harmonic semicircle previously constructed; thus, project points horizontally to the corresponding cam angles. Leave space for the 140-deg dwell.

STEP 6. Complete the displacement diagram with a French curve.

Before the cam is laid out, let us take a careful look at the nomogram sketch shown in Fig. 6.26(b). If the cam center is located within or on the boundary of the shaded section, the pressure angles obtained will be less than or equal to those specified. The specified pressure angles are 30 deg for rise and 50 deg for fall. Thus the minimum cam size which meets the stated requirements is at the apex O_1. Since the offset associated with O_1 is small, it would be better to use a radial follower design; thus we shall use point O_3 as the cam center. The maximum pressure angles associated with O_3 are, 21.8 deg for rise and 50 deg for fall. The prime radius $R_a = O_3 R(K) = 1.5(1.5) = 2.25$ in. The base radius $R_b = R_a - R_r = 2.25 - .5 = 1.75$ in. The completed cam is shown in Fig. 6.26(c); refer to Sect. 6.5 for the construction procedure.

6.13 Undercutting

Undercutting is a condition where the output motion from a cam-follower mechanism does not agree with the desired motion. Such a condition is shown in Fig. 6.27; here the cam surface cannot be drawn tangent to follower position 2. Consequently, the follower motion at position 2 will not be in accord with the planned motion.

Undercutting can be checked from a cam layout. If the cam surface cannot be drawn to each follower position, we have a case of undercutting.

For a plate cam operating with a radial roller follower, it is possible to check for undercutting by using the charts shown in Figs. 6.28 and 6.29. Example 6.4 illustrates how the charts are used. For proper action between a cam and a

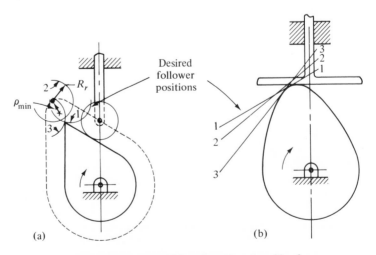

Fig. 6.27. Cams exhibit undercutting at position 2.

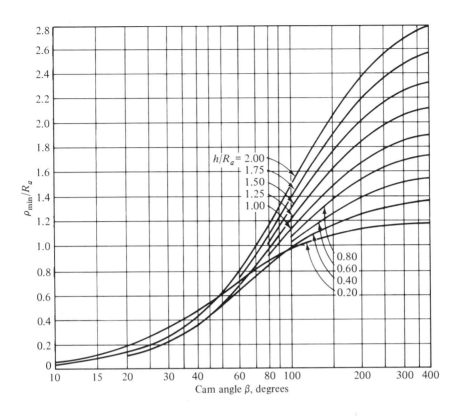

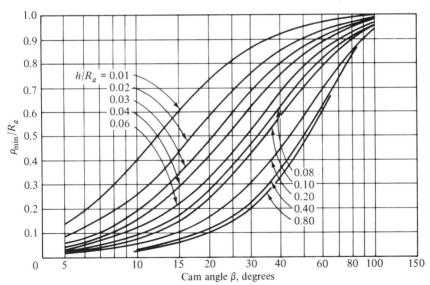

Fig. 6.28. Simple harmonic motion, undercutting chart for radial roller followers. (Courtesy of Product Engineering. From "Plate Cam Design—Radius of Curvature," M. Kloomok and R. V. Muffley, Sept. 1955.)

215

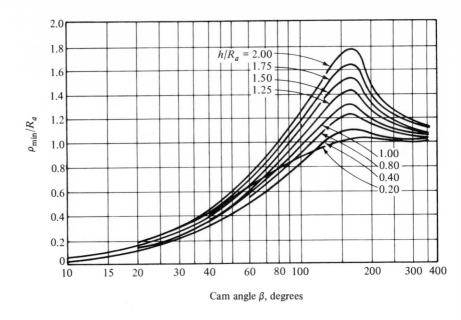

Cam angle β, degrees

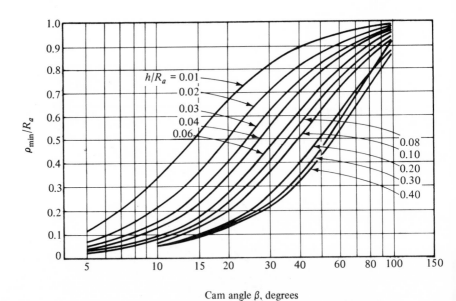

Cam angle β, degrees

Fig. 6.29. Cycloidal motion, undercutting chart for radial roller followers. (Courtesy of Product Engineering. From "Plate Cam Design—Radius of Curvature," M. Kloomok and R. V. Muffley, Sept. 1955.)

216

radial roller follower (that is, for no undercutting), $\rho_{min} > R_r$, where ρ_{min} is the minimum radius of curvature of the pitch curve and R_r is the radius of the roller. It can be seen in Fig. 6.27(a) that when ρ_{min} (position 2) is smaller than R_r, the desired position for the roller cannot touch the cam surface; thus, undercutting is present. When using the undercutting charts, check for undercutting for both the rise and the fall portions.

If undercutting is present, there are several ways of eliminating this condition:

1. Increase the prime radius, R_a.
2. Decrease the roller radius, R_r.
3. Increase the cam angle, β.
4. Decrease the maximum follower displacement, h.

Since several interrelating factors are present here, it is suggested that after a change or changes are made, the design be rechecked for undercutting.

EXAMPLE 6.4

Given: A radial roller follower moves through a total displacement of 1.25 in. Follower rises with cycloidal motion in 120 deg, dwells 110 deg, and then falls with cycloidal motion in 130 deg. The follower diameter $= 1$ in. The pressure angle for the rise portion should not exceed 35 deg.

Determine: Cam size (R_a) and check for undercutting.

Solution

STEP 1. For the cam size, use cycloidal nomogram Fig. 6.25. $R_a = .9(1.25) = 1.13$ in.; use 1.20 in.

STEP 2. For undercutting, refer to cycloidal chart Fig. 6.29. Calculate $h/R_a = 1.25/1.20 = 1.04$. For the rise portion, locate $\beta = 120$ deg and curve $h/R_a = 1.04$; since 1.04 is not on the chart, use $h/R_a = 1.00$. Obtain, approximately, $\rho_{min}/R_a = 1.1$.

STEP 3. Compare ρ_{min} to R_r. Since $\rho_{min}/R_a = 1.1$, $\rho_{min} = 1.1(R_a) = 1.1(1.2) = 1.3$ in. There is no undercutting here because $\rho_{min} > R_r$; numerically, 1.3 in. $> .5$ in.

The cycloidal fall portion was not checked for undercutting because $\beta = 130$ deg would result in a ρ_{min} larger than 1.3 in. If the values for ρ_{min} and R_r are close, it is advisable to avoid this marginal condition by increasing R_a, decreasing R_r, etc.; refer to the list of ways to eliminate undercutting.

6.14 Cam Fabrication

There are many methods of manufacturing a cam. Most of these methods can be grouped as cutting, grinding, forging, stamping, casting, and molding. Also, there are many materials to choose from—to name a few, steel, cast iron, plastic, and powdered metal. Some materials are limited to a specific manufacturing process. For example, a plastic cam is usually manufactured on a plastic molding machine. Factors which enter into the decision as to what material and manufac-

Fig. 6.30. Four-axis numerically controlled contouring machine. Cutter is controlled directly by a punched tape. This tape can be prepared by a digital computer. The tape reader is shown on the left side of the photograph. (Courtesy of Colt Industries Pratt and Whitney, Inc., Machine Tool Div.)

Fig. 6.31. Numerically controlled grinding of a hardened cam. Grinder is equipped with a General Electric Mark Century Control and Cam Technology Control. Generating accuracy is less than .0001 in. (Courtesy of Moore Special Tool Co., Inc.)

218

turing method to use are cam strength, accuracy, cost, environmental conditions, hardness, and durability. For more information, refer to *Product Engineering*, H. A. Rothbart, "Which Way to Make A Cam," March 1958, and M. A. Sanders, "When Plastic Cams Are Better," Aug. 1960.

Sometimes a cam is manufactured by carefully scribing the cam contour onto a steel plate. Then the plate is skillfully cut and hand finished to the proper shape. Obviously the final accuracy is dependent on the skill of both the draftsman and the machinist. This method is suitable for low-speed applications. A master cam is sometimes made in this fashion.

A few of the machines employed in the manufacture of cams are shown in Figs. 6.30 to 6.32. Figure 6.33 shows a cam-inspection device.

EXAMPLE 6.5

Given: A radial roller follower moves through a total displacement of .5 in. The follower rises with SHM in 90 deg, dwells 90 deg, then falls with SHM in 120 deg, and

Fig. 6.32. A 10-in.-diameter cylindrical cam is being cut on a commercial milling machine. This machine operates on the trace-milling principle, using a master as a model. The master or former cam is shown at the lower left. (Courtesy of American Cam Company, Inc.)

Fig. 6.33. Cam-inspection device will automatically inspect and read out actual displacements or compare a cam to a theoretical displacement tape, print out the error, and red-line a graph. It features radial increments of .001 deg or .1 min and displacement accuracy of plus or minus .00001 in. (Courtesy of Eonic, Inc.)

finally dwells for 60 deg. The base circle diameter = 1.25 in., the roller follower diameter = .50 in., and the cam rotates cw.

Determine: Rectangular cutter coordinates required to cut a plate cam. Assume that the cutter diameter is equal to the roller diameter of .50 in. Also, coordinates are to be accurate to one-thousandth of an inch.

Solution. The relationship between the polar and rectangular coordinates is shown in Fig. 6.34(a). The radial distance H from the cam center to the roller follower center is

$$H = R_a + y \tag{6.7}$$

where R_a is the prime radius and y is the follower displacement along the radial line. Assume that the roller follower and cutter are the same size. Then, rectangular coordinates for the cutter center (or roller center) are

$$X_c = H \cos \theta \tag{6.8}$$

$$Y_c = H \sin \theta \tag{6.9}$$

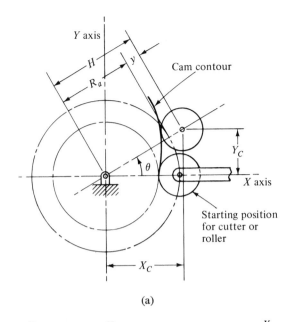

(a)

θ (deg)	X_C (in.)	Y_C (in.)	θ (deg)	X_C (in.)	Y_C (in.)
.00	.875	0.000	160.00	−1.292	.470
.75	.875	0.001	170.00	−1.354	.239
1.50	.875	0.023	180.00	−1.375	.000
2.25	.875	0.034	181.00	−1.375	−.024
3.00	.875	0.046	182.00	−1.374	−.048
3.75	.875	0.057	183.00	−1.372	−.072
4.50	.875	0.069	184.00	−1.370	−.096
5.25	.875	0.080	185.00	−1.368	−.120
6.00	.876	0.092	186.00	−1.364	−.143
6.75	.876	0.104	187.00	−1.361	−.167
7.50	.876	0.115	188.00	−1.356	−.191
8.15	.876	0.127	189.00	−1.351	−.214
9.00	.876	0.139	190.00	−1.346	−.237
18.00	.878	0.285	200.00	−1.261	−.459
27.00	.871	0.444	210.00	−1.127	−.651
36.00	.848	0.616	220.00	−.958	−.803
45.00	.795	0.795	230.00	−.765	−.911
54.00	.707	0.972	240.00	−.563	−.974
63.00	.577	1.133	250.00	−.363	−.996
72.00	.410	1.262	260.00	−.174	−.985
81.00	.213	1.346	270.00	.000	−.948
90.00	.000	1.375	280.00	.158	−.895
100.00	−.239	1.354	290.00	.302	−.830
110.00	−.470	1.292	300.00	.438	−.758
120.00	−.688	1.191	310.00	.562	−.670
130.00	−.884	1.053	320.00	.670	−.562
140.00	−1.053	.884	330.00	.758	−.438
150.00	−1.191	.688	340.00	.822	−.299
			350.00	.862	−.152

(b)

Fig. 6.34. (a) Cutter or follower coordinates for a radial roller follower cam. (b) Rectangular cutter coordinates X_C and Y_C, partial summary of results.

Angle θ is measured ccw from the X axis.

STEP 1. Determine the pressure angle and check for undercutting. Both are found to be satisfactory.

STEP 2. Determine the follower displacement y for SHM from the displacement table (Fig. 6.20). Use a .75-deg increment for the rise portion and a 1-deg increment for the fall portion.

STEP 3. Calculate values of H using Eq. (6.7) where $R_a = .625 + .250 = .875$ in.

STEP 4. Calculate cutter coordinates using Eqs. (6.8) and (6.9). A partial summary of the results is shown in Fig. 6.34(b).

The total solution to this problem requires using a .75-deg increment from 0 to 90 deg; thereafter a 1-deg increment is used. Calculating by hand would probably be overwhelming; therefore it is advisable to use either a desk calculator or preferably a digital computer. If a digital computer is used, Eqs. (6.3), (6.4), (6.7), (6.8), and (6.9) can be employed and the angular increment can be changed to a smaller value, say .50 deg from 0 to 360 deg. Many computers provide a punched tape output which can be directly used on a tape-controlled machine tool to manufacture the cam. For more information concerning cutter and cam equations, refer to Roger S. Hanson and Frederic T. Churchill, "New Cam Design Equations," *Product Engineering*, Aug. 1962.

PROBLEMS

6.1 Determine the displacement diagram for the eccentric cam shown in Fig. P6.1. Eccentricity and the roller diameter are each .5 in., and the cam diameter is 2 in. Use a 30-deg increment.

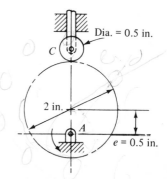

Fig. P6.1.

6.2 For the cam shown in Fig. P6.1, determine (a) the base circle radius, (b) the prime circle radius, (c) the maximum pressure angle for the rise and fall portions.

6.3 Design a translating cam similar to Fig. P6.3 which converts an x value to a y value. Where $y = \log x$, x can be any number between 100 and 1000.

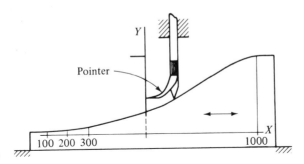

Fig. P6.3.

6.4 Design a timer cam to operate with a limit switch where the switch is off 64% of the cycle, on 15%, and then off 17%. Allow 2% of the cycle for each rise and fall motion. A .1-in. radial rise is required to actuate the switch. The base diameter is 1 in.

6.5 Lay out a plate cam (full size) which operates with a radial point follower. Start at the position shown in Fig. P6.5. The base radius is $1\frac{1}{4}$ in. Use the displacement diagram shown in Fig. 6.7. The cam rotates ccw.

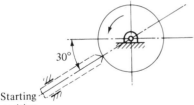

Fig. P6.5.   position

6.6 Lay out a plate cam (full size) which operates with a radial roller follower. Start at the position shown in Fig. P6.5. The base radius is $1\frac{1}{4}$ in., and the roller diameter is 1 in. Use the displacement diagram shown in Fig. 6.7. The cam rotates ccw.

6.7 Lay out a plate cam (full size) which operates with a radial flat-face follower. Start at the position shown in Fig. P6.5. The base radius is $1\frac{1}{4}$ in. Use the displacement diagram shown in Fig. 6.7. The cam rotates ccw. Determine the follower face length.

6.8 Lay out a plate cam (full size) which operates with an offset point follower. Start at the position shown in Fig. P6.8. The base radius is $1\frac{1}{4}$ in., and the offset is $\frac{5}{8}$ in. Use the displacement diagram shown in Fig. 6.7. The cam rotates ccw.

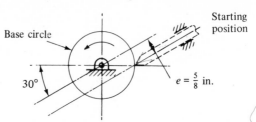

Fig. P6.8.

6.9 Lay out a plate cam (full size) which operates with an offset roller follower. Start at the position shown in Fig. P6.8. The base radius is $1\frac{1}{4}$ in., the prime radius is $1\frac{3}{4}$ in., and the offset is $\frac{5}{8}$ in. Use the displacement diagram shown in Fig. 6.7. The cam rotates ccw.

6.10 Lay out a plate cam (full size) which operates with an offset flat-face follower. Start at the position shown in Fig. P6.8. The base radius is $1\frac{1}{4}$ in., and the offset is $\frac{5}{8}$ in. Use the displacement diagram shown in Fig. 6.7. The cam rotates ccw. Determine the follower face length.

6.11 Lay out Prob. 6.5 using the overlay method. This method is outlined as follows: (1) Draw the base circle. Lay out all the follower positions at the starting position. (2) On a clear piece of vellum or tracing paper, locate a point O. Construct 30-deg radial lines through O. (3) Place the vellum over the base circle such that point O coincides with the center of the cam. Pin both sheets together at O. Rotate the vellum ccw every 30 deg, and each time mark the appropriate follower position on the vellum. When completed, draw in the cam contour.

6.12 Starting at the position shown in Fig. P6.12, lay out a plate cam (full size) which operates with the swinging roller follower. Use the displacement diagram shown in Fig. 6.7. Assume that this diagram is a plot of the follower position on the arc of motion vs. the cam angle. The cam rotates ccw.

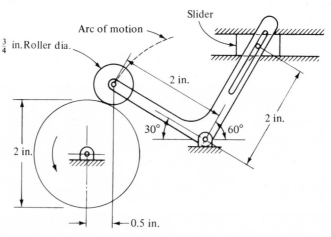

Fig. P6.12.

PROBLEMS

6.13 Starting at the position shown in Fig. P6.12, lay out a plate cam (full size) which operates with the swinging roller follower. Use the displacement diagram shown in Fig. 6.7. Assume that this diagram is a plot of the slider position vs. the cam angle. The cam rotates ccw.

6.14 Lay out a plate cam (full size) for Fig. P6.14. Use the displacement diagram shown in Fig. 6.7. Assume that this diagram is a plot of the follower position for point Q on the arc of motion vs. the cam angle. Q is on the follower face 4 in. from P. Determine face length L and distance X. The cam rotates ccw.

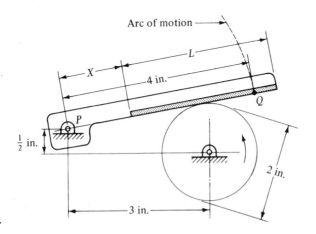

Fig. P6.14.

6.15 Lay out a plate cam (full size) for Fig. P6.14. Use the displacement diagram shown in Fig. 6.7. Assume that this diagram is a plot of the angular position of the follower vs. the cam angle. The total follower rise is 30 deg. The cam rotates ccw.

6.16 Starting from rest, a plate cam (Fig. P6.16) reaches an operating speed of 60 rpm in 4.5 sec. The equation of motion during the first 4.5 sec is $\theta = 40t^2$, where θ is the cam angle in degrees and t is time in seconds. (a) Graphically determine the position of the follower (displacement diagram) for the first 360 deg of cam rotation; use intervals of .2 sec. The cam contour was based on the displacement diagram shown in Fig. 6.7. (b) Compare the diagram determined in part a to that obtained when the cam is rotating at a constant 60 rpm.

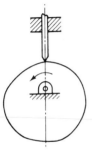

Fig. P6.16.

6.17 Construct a yoke cam (full size) which operates with radial-type flat-face followers. The base circle diameter is 2 in. Use the displacement diagram (0 to 180 deg) shown in Fig. 6.7. The cam rotates cw.

6.18 Construct a yoke cam (full size) which operates with radial-type roller followers. The base circle is 2 in., and the roller diameters are $\frac{3}{4}$ in. Use the displacement diagram (0 to 180 deg) shown in Fig. 6.7. The cam rotates cw.

6.19 Construct a displacement diagram to rise 2 in. with SHM in 120 deg of cam rotation, dwell 90 deg, fall 2 in. with SHM in 100 deg, and dwell for 50 deg.

6.20 Construct a displacement diagram to rise 2 in. with SHM in $\frac{1}{3}$ rev, dwell $\frac{1}{18}$ rev, fall 1 in. with SHM in $\frac{1}{4}$ rev, dwell $\frac{1}{18}$ rev, fall 1 in. with SHM in $\frac{1}{4}$ rev, and dwell for $\frac{1}{18}$ rev.

6.21 Construct a cam to operate with the translating circular follower shown in Fig. P6.21. The translating follower rises 1 in. with SHM in 120 deg of cam rotation, dwells 60 deg, falls with SHM in 150 deg, and then dwells for 30 deg. Is under-cutting present?

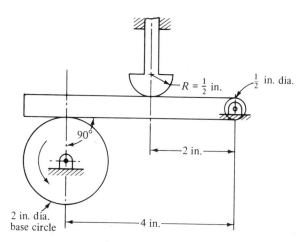

Fig. P6.21.

6.22 Using the displacement table (Fig. 6.20), construct a displacement diagram for the following program. Rise with SHM in 150 deg of cam rotation, dwell for 40 deg, fall with SHM in 110 deg, and then dwell for 60 deg. Use a relatively large increment for the cam angle. The maximum displacement is 1.5 in.

6.23 Construct a displacement diagram for Prob. 6.22 using the rise and fall equations for SHM.

6.24 Using the displacement table (Fig. 6.20), construct a displacement diagram for the following program. Rise with cycloidal motion in 150 deg of cam rotation, dwell for 40 deg, fall with cycloidal in 110 deg, and then dwell for 60 deg. Use a relatively large increment for the cam angle. The maximum displacement is 1.5 in.

6.25 Construct a displacement diagram for Prob. 6.24 using the rise and fall equations for cycloidal motion.

6.26 Ten translating cams are approximately equally spaced around a conveyor belt (see Fig. P6.26). An initial calculation shows the cam length to be approximately 2.157 in. Each cam drives the point follower. The point follower rises 1 in. with SHM for $\frac{3}{5}$ cam length, dwells for $\frac{1}{5}$ cam length, and then falls with SHM for $\frac{1}{5}$ cam length. Determine the cam contour. Use the displacement table (Fig. 6.20).

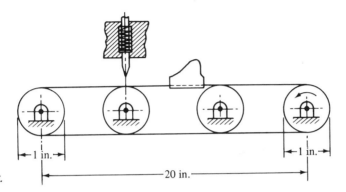

Fig. P6.26.

6.27 Determine the prime radius for a plate cam where the maximum follower displacement is .75 in. The radial follower dwells for 30 deg, rises with cycloidal motion in 80 deg, dwells for 50 deg, falls with SHM in 120 deg, and then dwells for 80 deg. The pressure angle for either the rise or fall portion must not be more than 35 deg.

6.28 A radial roller follower rises $\frac{1}{2}$ in. with SHM in 60 deg of cam rotation, dwells for 60 deg, again rises $\frac{1}{2}$ in. with SHM in 90 deg, dwells 30 deg, falls 1 in. with SHM in 90 deg, and finally dwells for 30 deg. The pressure angle is limited to 30 deg for the rise portions. Determine the prime radius for the plate cam.

6.29 A translating roller follower rises with cycloidal motion in 80 deg of cam rotation, dwells for 200 deg, falls with cycloidal motion in 35 deg, and then dwells for 45 deg. The maximum follower displacement is 1.25 in., the follower diameter is 1 in., and the cam rotates ccw. The pressure angle is limited to 35 deg for the rise portion and 45 deg for the fall portion. Determine the minimum cam size (prime radius).

6.30 A translating point follower rises with SHM for 100 deg of cam rotation, dwells for 100 deg, falls with SHM in 100 deg, and then dwells for 60 deg. The plate cam rotates cw, offset is .5 in., $R_b = 1.5$ in., and $h = 1$ in. Where would it be best to offset the follower, left or right of cam shaft? Why? Show numbers to prove answer.

6.31 In Fig. P6.31, a cam is to provide the following motions to a slider. The slider moves to the right 1 in. with cycloidal motion in 1 sec, dwells for 2 sec, returns with SHM in 2 sec, and finally dwells for 1 sec. This cycle occurs once for each cam rotation. The cam rotates at a constant rpm. A spring, not shown, holds the cam and roller together. Cycloidal motion is limited to a 30-deg pressure angle, and SHM is limited to a 45-deg pressure angle. Determine the minimum cam size (prime radius), base radius, rpm, direction of rotation, offset, and horizontal distance between cam center and roller center at the start of motion. Note that the relative position for the offset is shown in Fig. P6.31.

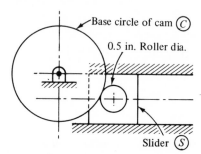

Fig. P6.31.

6.32 A radial point follower rises .5 in. with cycloidal motion in 25 deg of cam rotation, dwells for 100 deg, falls .5 in. with SHM in 100 deg, and finally dwells for 135 deg. The maximum pressure angle is restricted to 30 deg. Determine the prime radius for the plate cam.

6.33 Figure P6.33 shows a planetary cam mechanism. Input arm Ⓐ drives gear Ⓑ. The input arm is connected to Ⓑ at *P*. Rigidly fastened to gear Ⓑ is a follower arm. The follower arm carries a roller around a stationary cam track. As the roller and follower arm move about the cam, the follower arm rotates gear Ⓑ, which in turn rotates gear Ⓒ. Gear Ⓒ is connected to the output. Construct a cam for the input-output relationship shown in Fig. P6.33. The arm Ⓐ rotates ccw at a constant angular velocity. The diameters of Ⓑ and Ⓒ are 2 and 1 in., respectively. Distance *PQ* is 2 in., and the roller at *Q* has a diameter of $\frac{5}{16}$ in. Follower arm *PQ* is initially positioned 55 deg from the vertical.

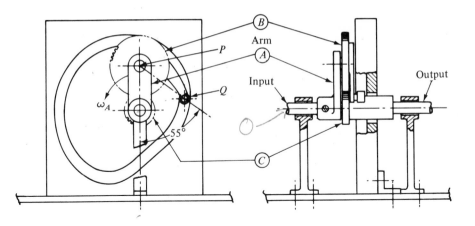

Input arm angle (deg)	Output gear angle (deg)		Input arm angle (deg)	Output gear angle (deg)
ccw	ccw		ccw	ccw
0.00	0.00		171.00	74.20
10.00	0.00		184.50	98.28
20.00	0.00		198.00	124.38
30.00	0.00		211.50	151.84
40.00	0.00		225.00	180.00
50.00	0.00		238.50	208.16
60.00	0.00		252.00	235.62
70.00	0.00		265.50	261.72
80.00	0.00		279.00	285.80
90.00	0.00		292.50	307.28
103.50	2.22		306.00	325.62
117.00	8.81		319.50	340.38
130.50	19.62		333.00	351.19
144.00	34.38		346.50	357.78
157.50	52.72		360.00	360.00

Fig. P6.33. (Courtesy of Addressograph Multigraph Corp.)

6.34 Make an input-output table for a planetary cam mechanism (Fig. P6.33), using the following data. When the input arm rotates from 0 to 90 deg, the output gear does not move; thus, 0 deg output. When the input arm rotates from 90 to 360 deg, the output gear moves with cycloidal motion. From 90 to 360 deg, use a 13.5 deg input increment. Note in Fig. P6.33 (table) the output gear exhibits SHM when the arm moves from 90 to 360 deg. For a description of the planetary cam mechanism refer to Prob. 6.33.

6.35 A radial roller follower rises 1.8 in. with SHM in 150 deg of cam rotation. The cam rotates at 420 rpm. (a) Determine the follower velocity at a cam angle of 30 deg. Use the velocity equation for SHM

$$V = \frac{h\pi\omega}{2\beta} \sin \frac{\pi\theta}{\beta}$$

where h = maximum displacement of follower (in.), β = cam angle corresponding to h (rad), θ = cam angle (rad), and ω = angular velocity (rad/min). (b) Determine the maximum follower velocity and the cam angle at which this occurs. (c) Determine the pressure angle at a cam angle of 30 deg. The pressure angle equation for a radial follower is

$$\tan \alpha = \frac{V}{(R_a + y)\omega}$$

where V = velocity (in./min), R_a = prime radius (in.), y = follower displacement corresponding to the cam angle θ (in.), and ω = angular velocity (rad/min). The prime radius is 2.5 in.

6.36 A radial roller follower rises with cycloidal motion in 100 deg, dwells for 10 deg, falls with SHM in 90 deg, and dwells for 160 deg. The roller radius is .5 in., and the total displacement is .5 in. Determine the cam size (prime radius) and check for undercutting. Modify cam if necessary.

6.37 Determine if undercutting is present; if so, modify the cam. A radial roller follower rises 1 in. with SHM in 50 deg of cam rotation, dwells 50 deg, falls with SHM in 80 deg, and dwells for 180 deg. The base radius is 1 in., and the roller radius is .625 in.

6.38 Determine if undercutting is present; if so, modify the cam. A radial flat-face follower rises .5 in. with SHM in 50 deg of cam rotation, dwells for 50 deg, falls with SHM in 180 deg of cam rotation, and then dwells for 80 deg. The base radius is 1 in.

6.39 Figure P6.39 shows a transfer mechanism which utilizes a stationary cam ⓒ. Four-bar linkage $ABCD$ is connected to disk ⓓ at A and D. Disk ⓓ is driven by an oscillating mechanism (not shown). As disk ⓓ rotates 90 deg cw about shaft center O from the position shown, link FAB moves 45 deg cw relative to the disk. Using SHM, determine the cam contour required to accomplish this motion. Linkages are on several different planes and do not interfere with one another.

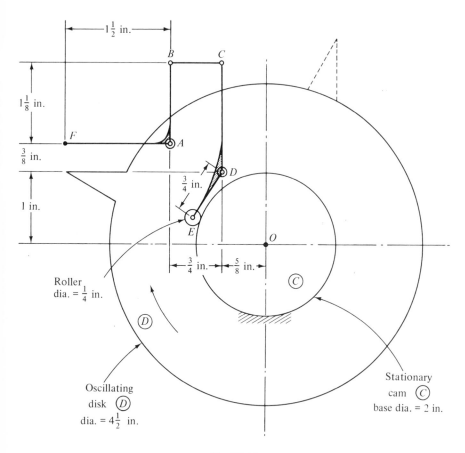

Roller dia. = $\frac{1}{4}$ in.

Oscillating disk $\textcircled{D}$ dia. = $4\frac{1}{2}$ in.

Stationary cam $\textcircled{C}$ base dia. = 2 in.

Fig. P6.39.

231

6.40 Figure P6.40 shows a cylindrical cam used to drive the leg of a walking doll. The cam diameter is 1 in., the width is $\frac{1}{2}$ in., and the follower diameter is $\frac{1}{8}$ in. The follower rises $\frac{1}{4}$ in. with SHM in 240 deg of cam rotation, dwells for 30 deg, and then returns with SHM in 90 deg. Lay out the cam (triple size). The cam rotates cw when looking down at the mechanism.

Fig. P6.40. (Courtesy of Mattel, Inc.)

Chapter **7**

GEARS

7.1 Introduction

A gear is a multicam or multitooth machine element used to transmit motion to another gear by direct contact. Usually mating gears operate at a constant angular velocity ratio. Theoretically, if the driver gear maintains a constant angular velocity, the follower gear operates at a constant angular velocity.

7.2 Classification of Gears

Gear pairs can be grouped according to the orientation of their shaft axes. Gear shafts may be parallel, intersecting, or skewed. Figure 7.1 shows some of the common types of gears used today. Many of the definitions used in this chapter are from AGMA (American Gear Manufactures Association) 112.04, "Standard Gear Nomenclature—Terms, Definitions, Symbols and Abbreviations."

Spur gears. Spur gears [Fig. 7.1(a)] are cylindrical in form and operate on parallel axes. Their teeth are straight and parallel to the axis.

233

a

b

c

d

e

Fig. 7.1. *Parallel-shaft gear arrangements:* (a) *spur gears,* (b) *parallel helical gears,* (c) *herringbone gears,* (d) *internal gear. Intersecting-shaft gear arrangements:* (e) *straight bevel gears,* (f) *spiral bevel gears,* (g) *face gear. Skewed-shaft gear arrangements:* (h) *hypoid gears,* (i) *crossed helical gears,* (j) *worm gear,* (k) *spiroid gears.* (Courtesy of Fairfield Manfacturing Co., The Fellows Gear Shaper Co., Gleason Works, Philadelphia Gear Corp., and Spiroid Division of Illinois Tool Works Inc.)

f

g

h

i

J

k

Fig. 7.1. (Cont.)

235

Parallel helical gears. Parallel helical gears [Fig. 7.1(b)] operate on parallel axes. When both are external, the helices are of opposite hand.

Herringbone gears. Herringbone gears [Fig. 7.1(c)] are similar to double-helical gears. Each gear has right- and left-hand helical teeth and the gears operate on parallel axes.

Internal gears. An internal gear [Fig. 7.1(d)] is shown in mesh with an external spur gear. The internal gear has teeth formed on the inner surface of the cylinder.

Straight bevel gears. Straight bevel gears [Fig. 7.1(e)] have straight tooth elements, which if extended would pass through the point of intersection of their axes.

Spiral bevel gears. Spiral bevel gears [Fig. 7.1(f)] have teeth that are curved and oblique.

Face gears. A face gear [Fig. 7.1(g)] is shown in mesh with a spur gear. A face gear has teeth on the end face of a disk.

Hypoid gears. Hypoid gears [Fig. 7.1(h)] are similar in form to bevel gears but operate on nonintersecting axes. Practically all hypoid gears have spiral teeth that are curved and oblique.

Crossed helical gears. Crossed helical gears, formerly called spiral gears [Fig. 7.1(i)], operate on crossed axes and may have teeth of the same hand or of opposite hand.

Worm and worm gears. A worm [Fig. 7.1(j)] is shown in mesh with a worm gear. A worm is a gear with one or more teeth in a form which resembles screw threads.

Spiroid gears. Spiroid gears [Fig. 7.1(k)] are right-angle nonintersecting axis gears. A face-type gear meshes with a constant lead-tapered pinion. Speed ratios range from 1:10 to 1:468 for a single mesh. Spiroid is a trade name of Illinois Tool Works Inc.

Fig. 7.2. Rack and spur gear. (Courtesy of the Fellows Gear Shaper Co.)

Rack gears. A rack and spur gear are shown in Fig. 7.2. A rack has straight teeth which are formed on a flat surface.

7.3 Involute Tooth Form

The most commonly used gear tooth shape is the involute. The involute profile has several characteristics which make it suitable for gear applications:

1. The angular velocity ratio of mating gears is constant and is unaffected by small changes in the center distance between shafts.
2. Power is efficiently transmitted.
3. Gears can be manufactured economically.
4. Different size gears of the same system will mesh properly with each other.

An involute (Fig. 7.3) is formed by the trace of point A (on the string) as the taut string unwinds from a fixed cylinder. The cylinder from which the involute is generated is called the base circle. The base circle of an involute gear is a fixed value, as the number of teeth on a gear is a fixed value.

An important property of an involute curve is that the center of curvature of any point on the involute curve is located at the tangent to the base circle. For example, the center of curvature for A_1 is at T_1.

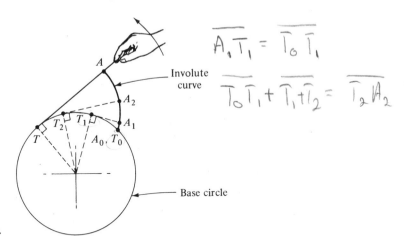

$$\overline{A_1 T_1} = \overline{T_0 T_1}$$

$$\overline{T_0 T_1} + \overline{T_1 T_2} = \overline{T_2 A_2}$$

Fig. 7.3. Involute curve.

7.4 Involute Action

Figure 7.4 shows two involute forms which are part of disks ⓓ and ⓕ. The involute on driver ⓓ rotates at a constant angular velocity about point A, driving the follower involute ccw about point B. It is assumed there is no impact at the start of contact. Involute curves are shown in contact at points X and Y, where X is on ⓓ and Y is on ⓕ. Points X and Y can be precisely

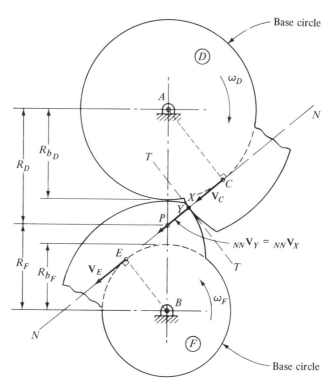

Fig. 7.4. *Involute action,* $\omega_F/\omega_D = R_D/R_F$.

established by drawing a normal line NN which passes through the center of curvature points C and E. The intersection of NN with the involutes locates X and Y. Note that normal line NN is tangent to the base circles. Regardless of where the contacting involutes are shown, normal line NN is always tangent to the base circles. Recall that normal velocity components of contacting points are equal. This is true for sliding as well as for pure rolling. Thus

$$_{NN}V_X = {}_{NN}V_Y$$

Utilizing the component method for points X and C on Ⓓ and points Y and E on Ⓕ, we have

$$V_C = {}_{NN}V_X$$

and

$$V_E = {}_{NN}V_Y$$

It follows that

$$V_C = V_E \tag{1}$$

because $_{NN}V_X = _{NN}V_Y$. The linear velocity of points C and E can be expressed as

$$V_C = R_{b_D}\omega_D \tag{2}$$

and

$$V_E = R_{b_F}\omega_F \tag{3}$$

where R_b is the base radius. Substitute Eqs. (2) and (3) into Eq. (1):

$$R_{b_D}\omega_D = R_{b_F}\omega_F$$

or

$$\frac{\omega_F}{\omega_D} = \frac{R_{b_D}}{R_{b_F}} \tag{4}$$

Right triangles APC and BPE are similar; thus the sides of the triangles are proportional. A proportional equation is

$$\frac{R_{b_D}}{R_{b_F}} = \frac{R_D}{R_F} \tag{5}$$

where R_D and R_F are referred to as the pitch radii and point P is called the pitch point. Substitution of Eq. (5) into Eq. (4) establishes the fundamental relationship between contacting involutes, namely

$$\frac{\omega_F}{\omega_D} = \frac{R_D}{R_F} \tag{7.1}$$

From the start of contact the involute surfaces continuously slide past one another except at pitch point P where pure rolling instantaneously exists. For proper involute action, involutes must contact along line NN. Changing the center distance between disks will not affect the involute action providing surfaces touch.

A series of properly spaced involutes on both disks Ⓓ and Ⓕ will allow Ⓕ to rotate completely at a constant angular velocity. These disks can now be called gears, and Eq. (7.1) is still applicable.

7.5 Spur Gear Terminology

Although this section is restricted to terminology concerning involute spur gears, many of the definitions can be extended to other types of gears. Terms not defined here are defined when required. Many of the following terms are illustrated in Figs. 7.5 and 7.6.

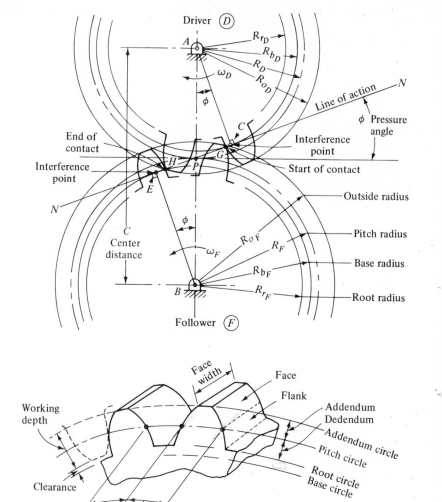

Fig. 7.5. Involute spur gear nomenclature.

Pitch circle. The pitch circle is an imaginary circle. When two gears are in mesh, the pitch circle of each gear roll together without slip.

Pitch diameter. The pitch diameter D (abbreviated PD) is the diameter of the pitch circle. The symbols D_D and D_F are used to distinguish between the pitch diameter of the driver and follower. Gear calculations are often based on the standard pitch diameter. The pitch radius is symbolized as R.

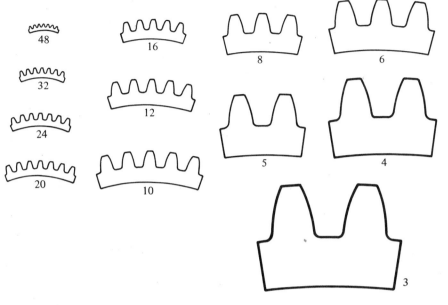

Fig. 7.6. Diametral pitch.

Number of teeth. N represents the number of teeth on a gear.

Circular pitch. The circular pitch p (abbreviated CP) is the distance measured along the pitch circle from a point on one tooth to the corresponding point on an adjacent tooth. Circular pitch equals the pitch circle circumference divided by the number of teeth on the gear:

$$p = \frac{\pi D}{N} \qquad (7.2)$$

Diametral pitch. Diametral pitch P (abbreviated DP) is the ratio of the number of teeth on a gear to the pitch diameter in inches:

$$P = \frac{N}{D} \qquad (7.3)$$

As a memory aid, N/D can be remembered as **N**umerator divided by **D**enominator. Frequently the diametral pitch is simply called pitch. A useful equation is obtained by multiplying Eq. (7.2) by Eq. (7.3):

$$p(P) = \frac{\pi D}{N}\left(\frac{N}{D}\right)$$

$$p(P) = \pi \qquad (7.4)$$

Diametral pitch is an index of the tooth size (Fig. 7.6). For example, a 1-in. pitch diameter gear with 20 teeth has a diametral pitch of $20/1 = 20$, whereas a 10-in. pitch diameter gear with the same number of teeth has a diametral pitch of $20/10 = 2$. Obviously the tooth size on the bigger gear is much larger than the tooth size on the smaller gear. Thus the smaller the diametral pitch, the larger the tooth size. Diametral pitch is usually a whole number.

Pitch point. The pitch point is the point of tangency of two pitch circles.

Pressure angle. The pressure angle ϕ is the angle between the normal line NN and the common tangent to the pitch circles. The standard pressure angles encountered are $14\frac{1}{2}$ deg, 20 deg, and 25 deg.

Outside radius. The outside radius R_o is the radius of the addendum circle. The addendum circle coincides with the top of the gear teeth.

Root radius. The root radius R_r is the radius of the root circle. The root circle coincides with the bottom of the tooth spaces.

Base radius. The base radius R_b is the radius of the base circle. Involute gear teeth are generated from the base circle. If the base radius is larger than the root radius, as in Fig. 7.5, the tooth profile from the base circle to the root circle is a noninvolute curve. Gear drawings often depict this curve as a radial line. Since

$$\sphericalangle PAC = \sphericalangle PBE = \phi$$

for both gears,

$$R_b = R \cos \phi \tag{7.5}$$

Length of action. The length of action is the distance that a pair of gear teeth are in contact. Figure 7.5 is specially drawn to show the start and end of contact for a pair of involute teeth. The initial engagement begins at G, where the flank of the driver tooth contacts the tip of the follower tooth. Contact proceeds along normal line NN to point H (addendum circle of driver), where contact ends. The length of action is $\overline{GH}$. The maximum length of action for involute teeth is the straight-line distance along normal line NN bounded by the addendum circles. Undercutting teeth and noninvolute action between mating teeth can reduce the length of action, as discussed in Sect. 7.9.

Line of action. Normal line NN is frequently called the line of action. The line of action is tangent to the base circles.

Interference points. Tangent points C and E on the base circles are called interference points.

Addendum. The addendum is the radial distance between the pitch circle and the addendum circle.

Dedendum. The dedendum is the radial distance between the pitch circle and the root circle.

Clearance. Clearance is the amount by which the dedendum of a gear exceeds the addendum of its mating gear.

Working depth. The working depth is the depth of engagement of two gears, that is, the sum of their addendums.

Tooth thickness. Tooth thickness or circular thickness is the length of arc between the two sides of a gear tooth measured on the pitch circle.

Tooth space. Tooth space is the length of arc between adjacent teeth measured on the pitch circle.

Face width. The face width is the length of a gear tooth measured parallel to the axis of rotation.

Face. Face is the surface area of a tooth between the pitch circle and addendum circle.

Flank. Flank is the surface area of a tooth between the pitch circle and root circle.

Backlash. Backlash B is the amount by which the width of a tooth space exceeds the thickness of the engaging tooth on the operating pitch circles.

Pinion. The smaller of two gears in mesh is often called the pinion. The larger gear is referred to as the gear.

Gear ratio. The gear ratio m_G is the ratio of the number of teeth on a gear to the number of teeth on a pinion. A gear ratio must be 1 or larger than 1.

Center distance. The center distance C (abbreviated CD) is the distance between gear axes:

$$C = R_D + R_F \tag{7.6}$$

Double both sides of Eq. (7.6):

$$2C = D_D + D_F \tag{1}$$

Since $P = N/D$ [Eq. (7.3)], $D = N/P$. Substituting for pitch diameters in Eq. (1), we obtain

$$2C = \frac{N_D}{P} + \frac{N_F}{P} \quad \text{or} \quad C = \frac{N_D + N_F}{2P} \tag{7.7}$$

7.6 Speed Ratio

The speed ratio or velocity ratio for a pair of gears is defined as the ratio of the angular velocity of the follower to the angular velocity of the driver, ω_F/ω_D. Beware, other literature may define the speed ratio as ω_D/ω_F.

Section 7.4, Eq. (7.1), showed that the speed ratio is inversely proportional to the pitch radii or pitch diameters:

$$\frac{\omega_F}{\omega_D} = \frac{D_D}{D_F} \tag{1}$$

The speed ratio can be expressed in terms of teeth number. Substituting for diameters in Eq. (1), where $D = N/P$ [Eq. (7.3)], we obtain

$$\frac{\omega_F}{\omega_D} = \frac{N_D/P}{N_F/P} \tag{2}$$

For gears to mesh properly, gear teeth must be of the same size; that is, both gears must have the same diametral pitch P. Equation (2) is reduced to

$$\frac{\omega_F}{\omega_D} = \frac{N_D}{N_F} \quad \leftarrow \text{No. of teeth} \tag{3}$$

Frequently the speed ratio for a gear pair is given as $d:f$; this means that

$$\frac{\omega_F}{\omega_D} = \frac{d}{f} \tag{4}$$

If $\omega_F = 100$ rpm and $\omega_D = 250$ rpm, then $d:f$ may be expressed as $2:5$. In ratio form this is

$$\frac{100}{250} = \frac{2}{5}$$

In summary, the speed ratio can be written as

$$\frac{\omega_F}{\omega_D} = \frac{N_D}{N_F} = \frac{D_D}{D_F} = \frac{R_D}{R_F} = \frac{d}{f} \tag{7.8}$$

Typical units are

$$\omega = \text{angular velocity, rpm}$$
$$N = \text{number of teeth}$$

$$D = \text{pitch diameter, in.}$$
$$R = \text{pitch radius, in.}$$
$$d, f = \text{dimensionless}$$

7.7 Pitch Radius

A problem which frequently occurs in spur gear design is to determine the pitch radii or diameters of a gear pair knowing the center distance C and the speed ratio $d:f$. Starting from the center distance equation [Eq. (7.6)],

$$C = R_D + R_F$$

divide both sides by $R_D + R_F$, multiply both sides by R_D, and rearrange:

$$R_D = C\left(\frac{R_D}{R_D + R_F}\right) \tag{1}$$

The speed ratio equation is

$$\frac{\omega_F}{\omega_D} = \frac{N_D}{N_F} = \frac{R_D}{R_F} = \frac{d}{f}$$

where $R_D = d(K)$ and $R_F = f(K)$; K is a particular number. Eq. (1) can now be expressed as

$$R_D = C\left(\frac{d(K)}{d(K) + f(K)}\right)$$

which simplifies to

$$R_D = C\left(\frac{d}{d + f}\right)$$

In the same manner it can be shown that

$$R_D = C\left(\frac{d}{d + f}\right) = C\left(\frac{N_D}{N_D + N_F}\right) = C\left(\frac{\omega_F}{\omega_D + \omega_F}\right) \tag{7.9}$$

To determine R_F, use

$$R_F = C\left(\frac{f}{d + f}\right) = C\left(\frac{N_F}{N_D + N_F}\right) = C\left(\frac{\omega_D}{\omega_D + \omega_F}\right)$$

or

$$R_F = C - R_D$$

or

$$R_F = \frac{f}{d}(R_D)$$

EXAMPLE 7.1

Given: Gear ratio for a pair of spur gears is 4:1. Pinion, which is the driver, has 16 teeth. Gear shafts are 5 in. apart. Driver rotates at 1800 rpm.

Determine: (a) number of teeth on the follower, (b) follower rpm, (c) pitch diameters, (d) diametral pitch, (e) circular pitch.

Solution. The completed sketch for this example is shown in Fig. 7.7.
(a) Recall gear ratio m_G was defined as

$$m_G = \frac{\text{teeth on gear}}{\text{teeth on pinion}}$$

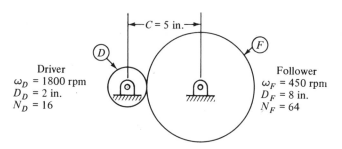

Driver
$\omega_D = 1800$ rpm
$D_D = 2$ in.
$N_D = 16$

Follower
$\omega_F = 450$ rpm
$D_F = 8$ in.
$N_F = 64$

Fig. 7.7. Completed sketch for Example 7.1.

In this example the pinion, which is the driver, has 16 teeth; thus

$$m_G = \frac{4}{1} = \frac{N_F}{16}$$

$$N_F = 4(16) = 64 \text{ teeth}$$

(b) Determine ω_F from the speed ratio equation [Eq. (7.8)]:

$$\frac{\omega_F}{\omega_D} = \frac{N_D}{N_F}$$

where $N_D = 16$ and $N_F = 64$. Solving for ω_F,

$$\omega_F = 1800\left(\frac{16}{64}\right) = 450 \text{ rpm}$$

(c) Use Eq. (7.9), slightly modified, to solve for D_D:

$$D_D = 2C\left(\frac{N_D}{N_D + N_F}\right) = 2(5)\left(\frac{16}{16 + 64}\right) = 2 \text{ in.}$$

and

$$D_F = 2C - D_D = 2(5) - 2 = 8 \text{ in.}$$

(d) Use Eq. (7.3) to determine P:

$$P = \frac{N_F}{D_F} = \frac{64}{8 \text{ in.}} = 8 \qquad \text{or} \qquad P = \frac{N_D}{D_D} = \frac{16}{2 \text{ in.}} = 8$$

(e) To determine circular pitch p, use Eq. (7.4):

$$p = \frac{\pi}{P} = \frac{\pi}{8} = .3927 \text{ in.}$$

7.8 Contact Ratio

Contact ratio m_C indicates the average number of tooth pairs in contact. A contact ratio of 1 means that one pair of teeth is in contact at any given instant. Owing to wear, machining error, tooth deflection under load, etc., a ratio of 1 would probably result in noisy operation. A ratio of less than 1 signifies that at times no teeth are in contact. This condition is unacceptable. For satisfactory service the minimum contact ratio most quoted is 1.4.

There are several ways of expressing the contact ratio for spur gears. The simplest is

$$m_C = \frac{\text{length of action}}{p \cos \phi} \tag{7.10}$$

where the length of action and circular pitch p are expressed in the same units. The length of action can be measured from a graphical layout (as in Example 7.2). For gears operating at a standard center distance the contact ratio can be quickly determined from the contact ratio-undercutting diagrams shown in, Baumeister and Marks, *Standard Handbook for Mechanical Engineers*, 7th ed. (McGraw-Hill, Inc., New York, 1967).

7.9 Interference

Interference occurs when the tip of one tooth, of a pair of contacting teeth, contacts or digs into the noninvolute flank of the mating tooth. Recall that the noninvolute portion of a tooth, if it exists, lies between the base and root circles.

The most severe case of interference is between a rack and spur gear of low tooth number. A rack (Fig. 7.8) is a gear of infinite radius where the involute tooth profile is a straight line perpendicular to the line of action.

Figures 7.8 and 7.9 show a sequence of contacting tooth profiles which exhibit interference. Interference can be seen at positions 2, 3, and 4.

One method of eliminating interference is to cut out or undercut the interfering noninvolute profile. Unfortunately, undercutting considerably weakens the tooth. Further, the contact ratio is reduced because the length of action has been decreased. Other methods of eliminating interference are

1. Avoid using pinions with low tooth numbers, especially for $14\frac{1}{2}$-deg standard gears.
2. Decrease the gear addendum.
3. Increase the pressure angle.

Interference can be detected from a graphical layout which shows the addendum circles and the line of action. If the length of action GH is within the interference points C and E, as in Fig. 7.5, interference will not occur. In Figs. 7.8 and 7.9, GH is not within CE; thus, interference is present. Note that in Fig. 7.8 point E is at infinity. A borderline condition exists when G coincides with C or when H coincides with E. A further possibility would be G and H simultaneously coinciding with C and E.

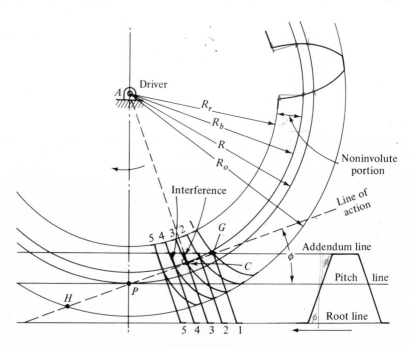

Fig. 7.8. Gear and rack. Interference at positions 2, 3 and 4.

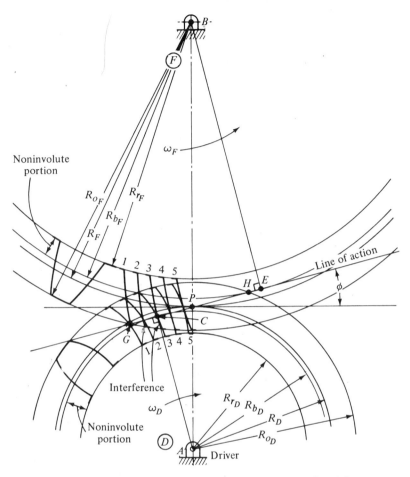

Fig. 7.9. Gear and pinion. Interference at positions 2, 3 and 4.

7.10 Standard Gear Systems

The term *standard* is often used to identify gear systems adopted by USASI*
(USA Standards Institute, successor to ASA) or AGMA. The most common
standard involute systems used today are the $14\frac{1}{2}$-deg, 20-deg, 25-deg full depth
and the 20-deg stub. Figure 7.10 lists the tooth proportions for these standard
gear systems. Other gear systems are also in use, particularly in the aircraft and
automotive industries.

Each system shown in Fig. 7.10 has its advantages and disadvantages. A
comparison between $14\frac{1}{2}$-, 20-, and 25-deg full depth systems shows the $14\frac{1}{2}$-deg
system to have a higher contact ratio. However, this can be offset by interference. Interference may occur for $14\frac{1}{2}$-deg gears when the pinion has less than 32

*USASI was changed to ANSI (American National Standards Institute, Inc.) on October
1969. ANSI is located at 1430 Broadway, New York, N.Y., 10018.

	$14\frac{1}{2}$ and 20 deg full depth obsolete (ASA B.6-1932)	20 deg stub obsolete (ASA B.6-1932)	20 and 25 deg full depth coarse pitch from $19.99P$ and coarser (AGMA 201.02)	$14\frac{1}{2}$, 20 and 25 deg full depth fine pitch from $20P$ and finer (AGMA 207.05)
Addendum	$\frac{1.000}{P}$	$\frac{0.800}{P}$	$\frac{1.000}{P}$	$\frac{1.000}{P}$
Dedendum	$\frac{1.157}{P}$	$\frac{1.000}{P}$	$\frac{1.250}{P}$	$\frac{1.200}{P} + 0.002$
Clearance	$\frac{0.157}{P}$	$\frac{0.200}{P}$	$\frac{0.250}{P}$	$\frac{0.200}{P} + 0.002$

Fig. 7.10. *Tooth proportions for involute spur gears. Standards are obtainable from AGMA, 1330 Massachusetts Avenue, N.W., Washington, D.C. 20005. AGMA 201.02 contains information pertaining to $14\frac{1}{2}$-deg. full depth and 20-deg. stub systems.*

teeth. Although the stub tooth profile is shorter than the full depth form, it is stronger than the full depth tooth. The stub system has a lower contact ratio than the full depth systems. Full depth 20-deg gear systems are widely accepted because they run smoother and quieter than the $14\frac{1}{2}$-deg system. A disadvantage of the 20-deg systems, compared to the $14\frac{1}{2}$-deg systems, is the force increase on the bearings. Gears manufactured from sintering and injection molding processes tend to use the 25-deg system. Fine pitch gears are primarily used for electromechanical applications, such as computers, timers, and servo systems, where tooth strength is generally not a critical factor.

For a pair of involute spur gears to operate properly together, each gear must be of the same

1. Gear system
2. Diametral pitch
3. Pressure angle

To avoid interference between an involute rack and pinion, the pinion should contain no fewer teeth than those indicated in Fig. 7.11. This table can be used as a guide for spur gear pairs. That is, if the pinion of a spur gear set contains fewer teeth than those listed in Fig. 7.11, interference is probably present. To check for interference, lay out the gear pair (as in Example 7.2) or refer to the contact ratio-undercutting diagrams shown in the *Standard Handbook for Mechanical Engineers* (*op. cit.*).

	20 deg Stub	$14\frac{1}{2}$ deg Full depth	20 deg	25 deg
Pinion Teeth	14	32	18	12

Fig. 7.11. Minimum number of teeth on a spur gear pinion to avoid interference with a rack.

EXAMPLE 7.2

Given: 4.000 in. and 14.000 in. pitch diameter spur gears were selected from a gear catalog. Driver has 24 teeth, follower 84 teeth. Gears are standard 20-deg full depth with a pitch of 6.

Determine: (a) Interference, if any; (b) contact ratio.

Solution. This problem can be solved without knowing the driver or the directions of rotation.

(a) Figure 7.11 shows that a 20-deg full depth pinion containing 18 teeth or more will not cause interference; therefore, interference is not present in this gear set since the pinion contains 24 teeth.

(b) The length of action will be determined graphically.

STEP 1. Refer to Fig. 7.12. Locate centers of rotation A and B 9 in. apart on the line of centers.

STEP 2. Locate pitch point P, 2 in. from A.

STEP 3. Through P draw a line perpendicular to the line of centers. This line is the common tangent to the pitch circles.

STEP 4. Draw the line of action through point P which makes a 20-deg angle with the common tangent line.

STEP 5. Calculate the addendum for a 20-deg full depth gear. From Fig. 7.10,

$$\text{Addendum} = \frac{1.000}{P} = \frac{1.000}{6} = .166 \text{ in.}$$

STEP 6. Add the addendum distance to each pitch radii and draw the addendum circles.

STEP 7. Label interference points C and E, and the beginning and end of contact G and H.

STEP 8. Measure length of action GH; $GH = .84$ in. The contact ratio is

$$m_C = \frac{\text{length of action}}{p \cos \phi}$$

where

$$p = \frac{\pi}{P}$$

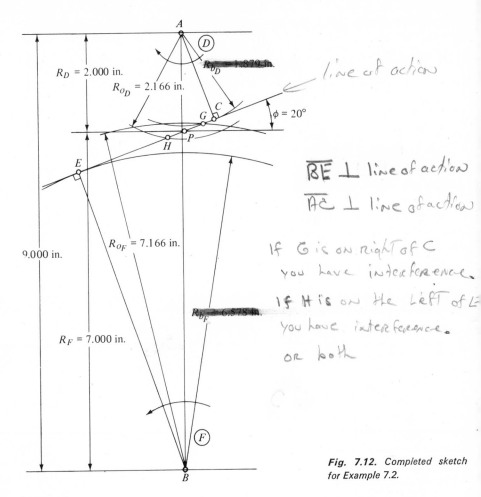

Fig. 7.12. Completed sketch for Example 7.2.

Therefore

$$m_C = \frac{.84}{(\pi/6)\cos 20°} = 1.71$$

The contact ratio is satisfactory because 1.71 is larger than 1.4.

Observe that GH is within CE, indicating, as predicted, no interference.

EXAMPLE 7.3

Given: Same information as Example 7.2 except gears are mounted on a 9.100 in. center distance.

Determine: (a) Pressure angle, (b) pitch, (c) contact ratio, (d) speed ratio.

Solution. The standard center distance should be 9.000 in. Changing the center distance alters the pitch radii, pressure angle, contact ratio, and diametral pitch. The base radii, outside radii, and speed ratio do not change.

STEP 1. Refer to Fig. 7.13. Locate centers of rotation A and B 9.100 in. apart on the line of centers.

STEP 2. Calculate the pitch radii. Using Eq. (7.9),

$$R_D = C\left(\frac{N_D}{N_D + N_F}\right) = 9.100\left(\frac{24}{24 + 84}\right) = 2.022 \text{ in.}$$

$R_F = C - R_D = 9.100 - 2.022 = 7.078$ in.

STEP 3. Locate pitch point P 2.022 in. from A.

STEP 4. Calculate the base radii. (The base radii were not required for Example 7.2 because the pressure angle 20 deg was known.)

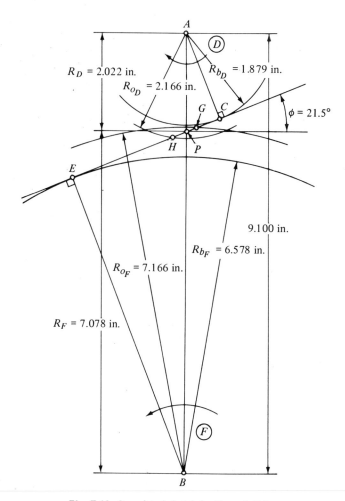

Fig. 7.13. Completed sketch for Example 7.3.

$$R_{b_D} = R_D \cos 20° = 2.000(.9397) = 1.879 \text{ in.}$$
$$R_{b_F} = R_F \cos 20° = 7.000(.9397) = 6.578 \text{ in.}$$

STEP 5. Draw in the base circles.

STEP 6. Draw the line of action tangent to the base circles and passing through point P.

STEP 7. Draw the addendum circles where $R_{o_D} = 2.166$ in. and $R_{o_F} = 7.166$ in.

STEP 8. Label points C, E, G, and H.

(a) The pressure angle can be calculated from Eq. (7.5):

$$\cos \phi = \frac{R_{b_D}}{R_D} = \frac{1.879}{2.022} = .930$$
$$\phi = 21.5°$$

(b) Pitch:

$$P = \frac{N_D}{D_D} = \frac{24}{2(2.022)} = 5.93$$

(c) Measure GH; $GH = .55$ in. Contact ratio:

$$m_C = \frac{GH}{(\pi/P)\cos\phi} = \frac{.55}{(\pi/5.93).930} = 1.12$$

(d) Speed ratio:

$$\text{Speed ratio} = \frac{N_D}{N_F} = \frac{24}{84} = \frac{2}{7}$$

or

$$\text{Speed ratio} = \frac{R_D}{R_F} = \frac{2.022}{7.078} \approx \frac{2}{7}$$

Note the speed ratio is more accurately determined from the tooth ratio.

7.11 Design of a Spur Gear Pair

One of the most important questions concerning a power gear pair is what horsepower can be safely transmitted. For a gear pair where size and cost are of primary concern, a detailed "strength" analysis is necessary. This analysis usually requires a trial-and-error procedure which involves such factors as tooth strength, surface durability, pitch-line velocity, tooth precision, and lubrication. Analysis of this type falls into the realm of machine design. For further information regarding this procedure, refer to any standard machine design text.

Strength and kinematic properties are interrelated in the design of a gear pair. For example, the diametral pitch or tooth size is related to the load-carrying capacity of a gear; from a kinematic standpoint, $P = N/D$. Solving the kinematic equation $P = N/D$ often requires an assumption about the diametral pitch. In this text either a diametral pitch is given (or can be easily calculated) or a range of acceptable diametral pitches is stated. It must be understood that the diametral pitch must satisfy the strength requirements.

There are many applications where a gear pair can be overdesigned. For these cases stocked gears from a gear catalog may be used. The procedure for selecting gears from a catalog is a simple one. The *Boston Gear Catalog* lists the approximate horsepower a gear carries when operating at a particular rpm. The *American Stock Gear Catalog* uses the equation

$$HP.F. = \frac{HP}{(RPM/100)(M.F.)(L.F.)(S.F.)}$$

where HP.F. = horsepower factor, horsepower rated at 100 rpm
$\quad$ HP = horsepower transmitted at desired rpm
$\quad$ RPM = revolutions per minute
$\quad$ M.F. = modifying factor
$\quad$ L.F. = lubrication factor
$\quad$ S.F. = service factor

Owing to the limited number of gears listed in a catalog, it is possible that the appropriate gears cannot be found. In this event the gear set must be designed.

Commercially packaged gears which are classified as speed reducers or increasers are available. They are obtainable in various sizes, ranging from a fraction of a horsepower to hundreds of horsepower.

The rpm of a gear is limited by load, tooth accuracy, and mass unbalance. As the rpm increases, impact between mating teeth increases. To reduce this effect, limitations are placed upon the pitch-line velocity. The pitch-line velocity is the speed of any point on the pitch circle. To calculate the pitch-line velocity, use the equation

$$V = \pi D n$$

where V = velocity, ft/min
$\quad D$ = pitch diameter, ft
$\quad n$ = rpm

Commercially cut (AGMA 390.02, Quality 5 or 6) metal spur gears are conservatively limited to a pitch-line velocity of 2000 ft/min. Fine pitch precision cut gears (AGMA 390.02, Quality 13 or 14) may operate as high as 10,000 ft/min.

When designing a gear pair, keep in mind the information shown in Figs. 7.14 and 7.15. Figure 7.14 represents typical values for speed ratios, efficiencies, and pitch-line velocities for various gears. Efficiency is defined as the ratio of output power to input power. Diametral pitches which are commonly used are shown in Fig. 7.15.

The face width for a spur gear commonly ranges between $10/P$ to $12/P$.

Type of gearing	Ratio range	Efficiency at rated power (%)	Maximum pitch-line velocity (ft/min)
Spur	1 - 10	98	2,000
Helical & herringbone	1 - 15	98	5,000
Helical and double			
helical, high speed	1 - 15	98	30,000
Crossed helical	1 - 10	98	4,000
Straight bevel	1 - 6	98	1,000
Spiral bevel	1 - 9	98	8,000
Hypoid	1 - 9	98	4,000
High-reduction hypoid	10 - 20	80	—
Single-enveloping worm	3½ - 90	50 - 90	6,000
Double-enveloping worm	3½ - 90	50 - 98	4,000
Face	3 - 8	95 - 99	4,000
Spiroid	10 - 100	50 - 97	6,000

Fig. 7.14. Comparison of single-mesh gears. Refer to S. L. Crawshaw and H. O. Kron, "Gears," Machine Design—Mechanical Drives Issue, Dec. 18, 1969. (Courtesy of Machine Design.) *Efficiency is based on speed-reducing units.*

Commonly used diametral pitches for spur gears					
Coarse			Fine		
$\frac{1}{2}$	3	12	20	64	128
1	4	14	24	72	150
2	6	16	32	80	180
2.25	8	18	40	96	200
2.5	10		48	120	

Fig. 7.15. Suggested diametral pitches for spur gears.

EXAMPLE 7.4

Given: Driver gear for a pair of standard 20-deg full depth spur gears rotates at 5000 rpm. Driven gear is to operate at approximately 1825 rpm. Center distance is approximately $1\frac{1}{4}$ in. Pitch-line velocity is limited to 2000 ft/min. Fine pitch application (between 32 and 80).

Determine: Pitch, pitch diameters, and teeth for a gear set which comes closest to an output of 1825 rpm. Check pitch-line velocity.

Solution. Before a gear set can be chosen, it is necessary to obtain the driver pitch diameter and speed ratio based on $\omega_F = 1825$ rpm and $C = 1\frac{1}{4}$ in. Using Eq. (7.9),

$$D_D = 2C\left(\frac{\omega_F}{\omega_D + \omega_F}\right) = 2(1.25)\left(\frac{1825}{5000 + 1825}\right) = 2.5\left(\frac{1825}{6825}\right) = .665 \text{ in.}$$

The speed ratio is

$$\frac{\omega_F}{\omega_D} = \frac{1825}{5000} = .3650$$

A check on the pitch-line velocity shows

$$V = \pi D_D n_D = \pi\left(\frac{.665}{12}\right)5000 = 870 \text{ ft/min}$$

This is satisfactory for spur gear design.

There are several restrictions placed upon the gear pair:

1. The speed ratio must be close to .3650, preferably exactly .3650.
2. The center distance must be close to $1\frac{1}{4}$ in. This means that the pitch diameter for the driver is approximately .665 in.
3. Diametral pitch is confined between 32 and 80.
4. Pitch-line velocity is restricted to a maximum of 2000 ft/min. (This was checked and found to be satisfactory.)

To simplify the presentation of the calculated results, the solution is laid out in tabular form in Fig. 7.16. The procedure is to assume a pitch P, which is

Assume P	$N_D = D_D P$	Whole N_D	$N_F = 2C(P) - N_D$	Whole N_F	$\dfrac{N_D}{N_F}$ = speed ratio
32	21.3 = 0.665(32)	21	59 = 2(1.25)32 − 21	59	$\frac{21}{59} = 0.356$
48	31.9 = 0.665(48)	32	88 = 2(1.25)48 − 32	88	$\frac{32}{88} = 0.364$
64	42.6 = 0.665(64)	43	117 = 2(1.25)64 − 43	117	$\frac{43}{117} = 0.368$
72	47.9 = 0.665(72)	48	132 = 2(1.25)72 − 48	132	$\frac{48}{132} = 0.364$
75	50 = 0.665(75)	50	137.5 = 2(1.25)75 − 50	137	$\frac{50}{137} \approx 0.365$

Fig. 7.16. *Trial-and-error solution for a spur gear pair.*

restricted to whole numbers, preferably the common ones shown in Fig. 7.15. Calculate the driver teeth based on the pitch diameter of $D_D = .665$ in. Round off N_D to the closest whole number of teeth. If the calculated value for N_D is exactly between two whole numbers, check out both numbers. Next, determine the number of teeth on the follower using the center distance equation [Eq. (7.7)]. Use a center distance of 1.25 in. Round off N_F to the closest whole number. Knowing N_D and N_F, determine the speed ratio N_D/N_F. Finally, after the table is completed, select a gear pair which has a speed ratio closest to .3650.

Referring to Fig. 7.16, of the commonly used diametral pitches 32,48, 64, and 72; 48 and 72 are the closest to the desired speed ratio of .3650. Let us select $P = 48$, $N_D = 32$, and $N_F = 88$. The pitch diameters are

$$D_D = \frac{N_D}{P} = \frac{32}{48} = .667 \text{ in.} \qquad \text{and} \qquad D_F = \frac{88}{48} = 1.834 \text{ in.}$$

$$C = \frac{N_D + N_F}{2P} = \frac{32 + 88}{2(48)} = 1.25 \text{ in.}$$

Note that the center distance was not altered because the calculated value for N_F (Fig. 7.16) is a whole number. Follower rpm is

$$\omega_F = \omega_D \frac{N_D}{N_F} = 5000 \left(\frac{32}{88} \right) \approx 1818 \text{ rpm}$$

A diametral pitch of 75 is illustrated in the table because this gear set is very close to .3650. A pitch of 75 is uncommon and therefore would cost more to manufacture, unless ordered in large quantities. For $P = 75$, $N_D = 50$, and $N_F = 137$:

$$D_D = \frac{50}{75} = .667 \text{ in.} \qquad \text{and} \qquad D_F = 1.827 \text{ in.}$$

$$C = \frac{50 + 137}{2(75)} = 1.247 \text{ in.}$$

Note that the center distance is not 1.25 in.

If a gear pair from a catalog is desired, it will probably be necessary to adjust the speed ratio. Use the gear catalog corresponding to $P = 32$, 48, 64, and 72. Select teeth numbers from the catalog which are close to those shown in Fig. 7.16. A slide rule or calculator can be quickly set up to show which gear pair is closest to .3650. For example, a selected pair for $P = 64$ may be $N_D = 44$ and $N_F = 120$, where $D_D = .688$ in. and $D_F = 1.875$ in.:

$$C = \frac{44 + 120}{2(64)} = 1.281 \text{ in.}$$

and

$$\omega_F = 5000\left(\frac{44}{120}\right) = 1839 \text{ rpm}$$

It should be realized that the ideal center distance must be increased to avoid the possibility of gears binding. As a numerical illustration, let us rather arbitrarily adjust an ideal center distance of 1.2813 to 1.2817 in., an increase of .0004 in. To maintain the minimum dimension of 1.2817 in., the center distance may be specified bilaterally as $1.2822 \pm .0005$ in. or unilaterally as $1.2817 \pm \genfrac{}{}{0pt}{}{.0010}{.0000}$ in. For more information on center distance tolerancing, refer to George W. Michalec, *Precision Gearing* (John Wiley & Sons, Inc., New York, 1966).

7.12 Helical and Herringbone Gears

A parallel helical gear pair [Fig. 7.1(b)] can transmit greater load at a higher pitch-line velocity and with less noise than a comparable spur gear pair. For this reason, helical and particularly herringbone gears [Fig. 7.1(c)] are used in high-speed heavy-duty applications. A helical gear pair can usually be designed to meet a specified speed ratio at a specified center distance. Thus there is more design flexibility with a helical pair than with a spur gear pair.

Most helical gear terminology is with reference to either the transverse or normal plane. The transverse plane is perpendicular to the axis of rotation. The normal plane is perpendicular to a helical tooth element. Subscripts t and n are used to identify terms in the transverse and normal planes. Some gear terms are illustrated in Figs. 7.17 and 7.18; these are simplified drawings of helical gears where the gears are sectioned at the pitch cylinder.

The helix angle ψ is measured between a tooth element and an axial element of the pitch cylinder. Circular pitches p_n and p_t are measured from tooth to tooth on the pitch cylinder; refer to Fig. 7.18, where

$$p_n = p_t \cos \psi \tag{1}$$

Diametral pitches P_n and P_t are related to the circular pitches

$$p_n = \frac{\pi}{P_n} \tag{2}$$

and

$$p_t = \frac{\pi}{P_t} \tag{3}$$

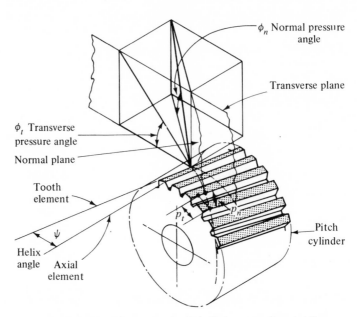

Fig. 7.17. *Simplified isometric of a helical gear, sectioned at the pitch cylinder. Shown are pressure angles, circular pitches, and helix angle.*

Substitution for p_n and p_t in Eq. (1) results in

$$P_t = P_n \cos \psi \tag{7.11}$$

This means that the normal diametral pitch P_n is larger than the transverse diametral pitch P_t. The spur gear relationship $P = N/D$ is also true for a helical gear in the transverse plane; thus

$$P_t = \frac{N}{D} \tag{7.12}$$

Using Eq. (7.11), substitute for P_t in Eq. (7.12):

$$P_n = \frac{N}{D \cos \psi} \tag{7.13}$$

The center distance for helical gears is

$$C = \frac{D_D}{2} + \frac{D_F}{2} = \frac{N_D}{2P_{t_D}} + \frac{N_F}{2P_{t_F}} = \frac{N_D}{2P_{n_D} \cos \psi_D} + \frac{N_F}{2P_{n_F} \cos \psi_F} \tag{4}$$

For a parallel pair of helical gears, the driver and follower have the same circular pitches ($p_{t_D} = p_{t_F}$ and $p_{n_D} = p_{n_F}$) and diametral pitches ($P_{t_D} = P_{t_F}$ and

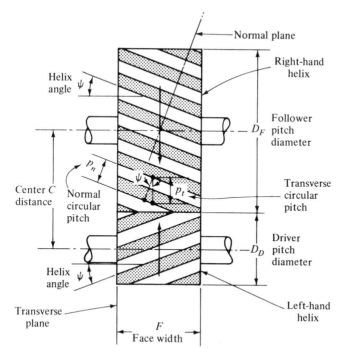

Fig. 7.18. Simplified drawing of a pair of helical gears mounted on parallel shafts. Gears are sectioned at the pitch cylinders.

$P_{n_D} = P_{n_F}$). Also, the helix angle of both gears are equal; however, they are of opposite hand. Equation (4) can be simplified to

$$C = \frac{N_D + N_F}{2P_t} = \frac{N_D + N_F}{2P_n \cos \psi} \tag{7.14}$$

The top gear in Fig. 7.18 is a right-hand helical gear. A right-hand helical gear can be identified in the following manner: with shaft axis horizontal, a right-hand helical gear slopes down to the right. A left-hand helix slopes down to the left.

The speed ratio for a helical pair looks identical to a spur gear pair, namely

$$\frac{\omega_F}{\omega_D} = \frac{D_D}{D_F} = \frac{N_D}{N_F}$$

To ensure overlapping tooth action, the minimum face width should be at least twice the axial pitch p_X; thus

$$\text{Minimum face width} = F = 2p_X = \frac{2p_t}{\tan \psi} = \frac{2\pi}{P_n \cos \psi \tan \psi}$$

The disadvantage of a helical gear is that it produces a thrust along the shaft. This thrust must be carried by the bearings. Figure 7.19 shows the rotation and axial direction of the thrust produced by a single pair of helical gears. To minimize the axial thrust, keep the helix angle low, between 7 and 23 deg. Commercial helical gears have helix angles of 45 deg. This type of gear can be used in parallel or crossed (90-deg) helical pairs.

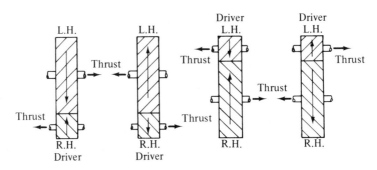

Fig. 7.19. Directions of rotation and thrust for helical gears mounted on parallel shafts. Thrust is from the helical gear onto the shaft.

Herringbone gears are used to transmit heavy loads. A herringbone gear produces equal and opposite axial forces, thus eliminating axial thrust.

Helical gears can be cut by either the hobbing or shaping method. (Sect. 7.19 discusses gear manufacturing.) Hobbing is preferable to shaping. If a gear is hobbed, the normal diametral pitch P_n and normal pressure angle ϕ_n are usually whole numbers, whereas with a shaped gear the transverse diametral pitch P_t and transverse pressure angle ϕ_t are usually whole numbers.

A helical pinion can operate satisfactorily, without interference, with less teeth than a spur gear pinion. Use the teeth number shown in Fig. 7.20 as a guide for helical gears. Note that the spur gear data shown in the table is repeated from Fig. 7.11.

EXAMPLE 7.5

Given: It is desired to reduce the speed of an electric motor using three gear pairs. The first gear pair (see Fig. 7.21) is to be a helical pair. The electric motor rotates at 900 rpm. Gear ratio is to be 9: 1. Center distance is .750 in. Normal pressure angle $\phi_n = 25$ deg, and normal diametral pitch is 72.

Determine: Pitch diameter, teeth, and helix angle for each helical gear. Also, determine a reasonable minimum face width.

Solution. Assuming $N_D = 10$ teeth, then $N_F = 90$ teeth. Helix angle is determined from Eq. (7.14):

Helix angle, ψ (deg)	Normal pressure angle, ϕ_n		
	$14\frac{1}{2}°$	$20°$	$25°$
0 (Spur Gear)	32	18	12
5	32	17	12
10	31	17	11
15	29	16	11
20	27	15	10
23	26	14	10
25	25	14	9
30	22	12	8
35	19	10	7
40	15	9	6
45	12	7	5

Fig. 7.20. *Minimum number of teeth on a helical pinion (of a parallel helical pair) to avoid interference. Extracted from AGMA 207.05.* (Courtesy of AGMA.)

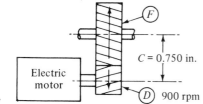

Fig 7.21.

$$\cos\psi = \frac{N_D + N_F}{2(P_n)C} = \frac{10 + 90}{2(72)(.750)} = .926$$

$$\psi = 22°11'$$

Pitch diameters are

$$D_D = 2C\left(\frac{d}{d+f}\right) = 2(.750)\left(\frac{1}{1+9}\right) = .150 \text{ in.}$$

and

$$D_F = 2C - D_D = 2(.750) - .150 = 1.350 \text{ in.}$$

A reasonable minimum face width based on kinematic requirements is

$$F = 2\left(\frac{\pi}{P_n \cos\psi \tan\psi}\right) = \frac{2\pi}{72 \cos 22°11' \tan 22°11'} = .231 \text{ in.}$$

7.13 Crossed Helical Gears

Helical gears that are connected between nonparallel and nonintersecting shafts are commonly called crossed helical gears [Fig. 7.1(i)]. Crossed helical gears are limited to light loads because the gear teeth theoretically contact at only one point.

The shaft angle, Σ, for a pair of crossed helical gears is either

$$\Sigma = \psi_D + \psi_F \qquad \text{[gears have same hand, Fig. 7.22(a)]}$$

or

$$\Sigma = |\psi_D - \psi_F| \qquad \text{[gears have opposite hand, Fig. 7.22(b)]}$$

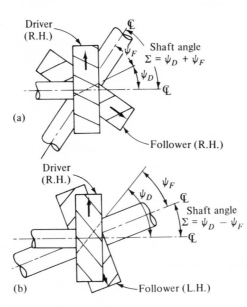

Fig. 7.22. Simplified sketch of crossed helical gear pairs. (a) Two right-hand helicals where the shaft angle $\Sigma = \psi_D + \psi_F$. (b) Right- and left-hand helicals where the shaft angle $\Sigma = \psi_D - \psi_F$. The helix angle is measured between the hidden dotted line, representing the helical tooth, and a line parallel to the shaft center line. To help visualize the follower motion, imagine the driver tooth (dotted line) pushing the follower tooth (dotted line.) Note that the arrow for the direction of rotation lies on top of the gear.

where ψ_D and ψ_F are the helix angles of the driver and follower. Most applications involve helicals of the same hand.

For crossed helical gears to mesh properly, each gear must have the same normal diametral pitch. The center distance equation can be developed from Eq. (4), Sect. 7.12 (page 260):

$$C = \frac{1}{2P_n}\left(\frac{N_D}{\cos\psi_D} + \frac{N_F}{\cos\psi_F}\right) \tag{7.15}$$

The speed ratio for crossed helical gears is

$$\frac{\omega_F}{\omega_D} = \frac{N_D}{N_F} \tag{1}$$

Substitute for N in Eq. (1); use Eq. (7.13), where $N = P_n D \cos \psi$:

$$\frac{\omega_F}{\omega_D} = \frac{P_{n_D} D_D \cos \psi_D}{P_{n_F} D_F \cos \psi_F} \qquad (2)$$

Since $P_{n_D} = P_{n_F}$,

$$\frac{\omega_F}{\omega_D} = \frac{N_D}{N_F} = \frac{D_D \cos \psi_D}{D_F \cos \psi_F} \qquad (7.16)$$

To reduce wear and increase efficiency, the helix angles of a mating pair should be equal or nearly equal.

EXAMPLE 7.6

Given: A crossed helical gear pair is to connect two shafts at a shaft angle of 90 deg. Gear ratio is 3.5: 1. Center distance is approximately 3 in. Normal diametral pitch is 12. Directions of rotation are shown in Fig. 7.23. Small gear is the driver.

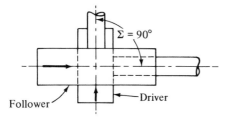

Fig. 7.23. Follower — Driver, $\Sigma = 90°$

Determine: Helix angle, number of teeth, pitch diameter, and hand of helix for each gear.

Solution. Change subscripts D and F in Eq. (7.15) to a more general form S and L, meaning small and large gear:

$$C = \frac{1}{2P_n}\left(\frac{N_S}{\cos \psi_S} + \frac{N_L}{\cos \psi_L}\right) \qquad (1)$$

Gear ratio $m_G = N_L/N_S$; thus Eq. (1) becomes

$$C = \frac{N_S}{2P_n}\left(\frac{1}{\cos \psi_S} + \frac{m_G}{\cos \psi_L}\right) \qquad (2)$$

A good starting point for crossed helical gears is to assume that the helix angles are equal; Eq. (2) simplifies to

$$C = \frac{N_S}{2P_n \cos \psi}(1 + m_G) \qquad (3)$$

STEP 1. Determine N_S from Eq. (3), where $\psi = \psi_S = \psi_L = 45$ deg, and $C = 3$ in.:

$$N_S = \frac{2CP_n \cos \psi}{1 + m_G} = \frac{2(3)(12) \cos 45°}{1 + 3.5} = 11.3 \text{ teeth}$$

Assume a smaller number of teeth than 11.3, say 10 teeth.

STEP 2. With $N_S = 10$, $N_L = m_G N_S = 3.5(10) = 35$ teeth. N_S and N_L must be whole numbers.

STEP 3. Now using $N_S = 10$, go back to Eq. (3) and solve for C:

$$C = \frac{N_S}{2P_n \cos \psi}(1 + m_G) = \frac{10}{2(12) \cos 45°}(1 + 3.5) = 2.652 \text{ in.}$$

STEP 4. The pitch diameters are

$$D_S = \frac{N_S}{P_n \cos \psi_S} = \frac{10}{12 \cos 45°} = 1.179 \text{ in.}$$

$$D_L = 2(2.652) - 1.179 = 4.125 \text{ in.}$$

STEP 5. For the rotations desired, both helicals should be left-hand; refer to Fig. 7.24.

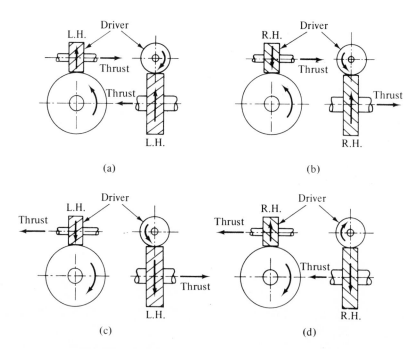

Fig. 7.24. *Directions of rotation and thrust for crossed helical gears mounted on right-angle shafts. Thrust is from the helical gear onto the shaft. This figure also applies to worm and worm gear pairs where the driver is considered as the worm.*

In summary,

Gear	D	N	ψ	Hand
Small (driver)	1.179	10	$45°$	Left
Large (follower)	4.125	35	$45°$	Left

Let us change Example 7.6 slightly. Suppose that the center distance must be kept at exactly 3.000 in. What is the solution now? Starting from the assumption that the helix angles are equal, we obtained, $N_S = 11.3$ teeth. Again assume that $N_S = 10$ teeth. Eq. (2) can be written as

$$\frac{2CP_n}{N_S} = \frac{1}{\cos \psi_S} + \frac{m_G}{\cos \psi_L}$$

Since $\Sigma = 90$ deg, $\cos \psi_L = \sin \psi_S$; therefore

$$\frac{2CP_n}{N_S} = \frac{1}{\cos \psi_S} + \frac{m_G}{\sin \psi_S} \quad \text{or} \quad \frac{2CP_n}{N_S} = \sec \psi_S + m_G \csc \psi_S$$

By a trial-and-error solution, we obtain

$$\frac{2(3)12}{10} = \sec 35°55.5' + 3.5 \csc 35°55.5'$$

$$7.200 = 1.234894 + 3.5(1.704374)$$

Gear	D	N	ψ	Hand
Small (driver)	1.029	10	$35°55.5'$	Left
Large (follower)	4.971	35	$54°4.5'$	Left

7.14 Worm and Worm Gear

A worm [Fig. 7.1(j)] can be considered as a special form of a crossed helical gear. Almost invariably the worm is the driver and the shaft angle is 90 deg. A worm resembles a screw thread, and for this reason there is similarity between worm and screw-thread terminology. Worm teeth are often called threads.

Worm gear drives can transmit more horsepower at larger speed reductions and with less noise than a crossed helical gear pair. Rubbing action between mating teeth causes a large frictional loss, and, as a result, efficiency is low. Typical range of efficiencies for worm gearing is shown in Fig. 7.14.

Sometimes there is confusion between the terms *worm* and *worm gear*. The worm is the pinion, whereas the large gear (or wheel) is called the worm gear. The lead angle λ is defined as the angle between a tooth or thread element (at the pitch diameter) and the plane of rotation (see Fig. 7.25). For right-angle

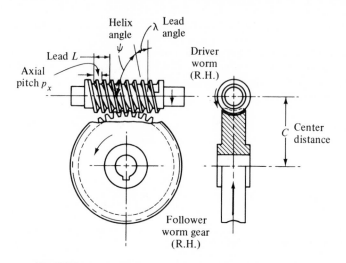

Fig. 7.25. Nomenclature for worm-worm gear. This is a single-enveloping worm. The worm is double-threaded; the lead is twice the axial pitch, $L = 2p_x$.

shafts the complement of the worm lead angle is the helix angle ψ. The axial pitch p_x of a worm (also called linear pitch) is the distance between corresponding points of adjacent threads parallel to the axis of rotation. Lead L, of a worm, is the distance that a point on the thread advances for 1 rev of the worm. For a single-threaded worm, the lead equals the axial pitch. For a double-threaded worm, $L = 2p_x$; for a triple-threaded worm, $L = 3p_x$; etc.

The most common class of worm gearing is the single-enveloping type (Fig. 7.25). Figure 7.26 shows the double-enveloping worm; both the worm and gear are throated. The double-enveloping pair can carry more load; however, the gearset must be accurately mounted.

A worm gearset is said to be self-locking when the worm gear cannot drive the worm. When the worm has a lead angle of less than 5 deg, the gearset is considered to be self-locking.

The speed ratio for a worm gear pair, assuming the worm to be the driver, is

$$\frac{\omega_F}{\omega_D} = \frac{N_D}{N_F} \tag{7.17}$$

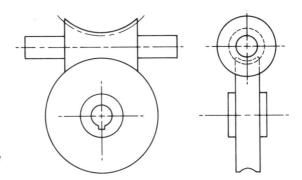

Fig. 7.26. Double-enveloping worm or cone worm set.

where ω_F = worm gear rpm (follower)

ω_D = worm rpm (driver)

N_D or thd = number of threads on the worm (as single, double, triple, etc.)

N_F = number of teeth on the worm gear

Use Fig. 7.24 to determine the directions of rotation and thrust for a worm or worm gear. Note that a meshing worm and worm gear have the same helix hand.

EXAMPLE 7.7

Given: A single reduction speed reducer (Fig. 7.27) contains a 30-tooth worm gear and a single-threaded worm. Input rpm is 1725.

Determine: Output rpm.

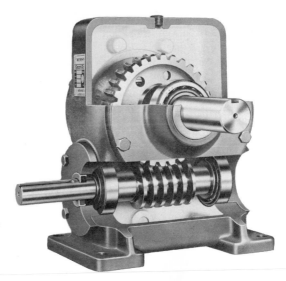

Fig. 7.27. Worm gear speed reducer. (Courtesy of Winsmith, Division of UMC Industries, Inc.)

Solution. Using Eq. (7.17),

$$\omega_F = \omega_D \left(\frac{N_D}{N_F}\right) = 1725\left(\frac{1}{30}\right) = 57.5 \text{ rpm}$$

7.15 Bevel Gears

Bevel gears are usually used to connect shafts at right angles. There are several types of bevel gears presently in use. To name a few, straight, coniflex, spiral, and hypoid.

The teeth on a straight bevel gear [Fig. 7.1(e)] are straight. If the teeth have been crowned, it is called a coniflex bevel gear. A crowned tooth (Fig. 7.28) is broadened at the center so that contact takes place at the center rather than at the ends. A coniflex gear pair produces quieter action.

Fig. 7.28. Typical shape of a crowned tooth.

A spiral bevel gear [Fig. 7.1(f)] has curved teeth. Compared to a straight bevel gear, the spiral bevel gear can carry more load quietly and at a higher pitch-line velocity. A spiral bevel pair produces thrust loads which must be opposed by the bearings. Spiral bevel gears can be left- or right-hand (Fig. 7.29). A spiral hand can be identified as follows: with the gear's axis of rotation horizontal, a left-hand spiral slopes down to the left, and a right-hand slopes down to the right.

Left-hand   Right-hand

Fig. 7.29. Simplified drawing of a left- and right-hand spiral bevel gear.

The speed ratio for a bevel pair is the same as for a spur gear pair, namely

$$\frac{\omega_F}{\omega_D} = \frac{N_D}{N_F}$$

Some bevel gear terminology is shown in Fig. 7.30.

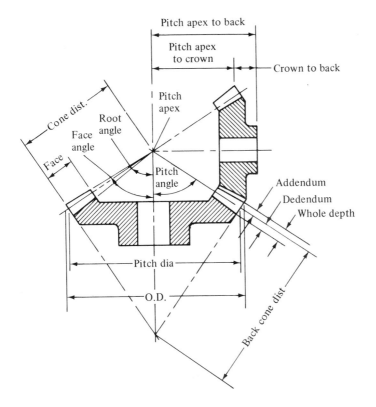

Fig. 7.30. *Bevel gear terminology.*

7.16 Backlash

To avoid a tight mesh between a gear pair and because of manufacturing tolerances, a gear manufacturer usually builds in backlash by thinning the teeth.

Backlash for a gear pair, as defined in Sect. 7.5, is the amount by which the width of a tooth space exceeds the thickness of the engaging tooth on the operating pitch.

For a gear pair or pairs, backlash can be determined by the following procedure. With the power off and the input shaft stationary, move the output shaft from the extreme clockwise position to the extreme counterclockwise position or vice versa. The angular movement of the output shaft is the backlash in the system. Apply only a small force to the shaft. It should be noted that this angular measurement does not involve the load and dynamic factors which affect backlash and further may not be representative of other positions of the gears.

Backlash usually must be minimized in instrument and servo design. If backlash is too high in a servo system, the system becomes unstable.

Precision spur gears are often used to minimize backlash. A precision gear is toleranced in thousandths and ten-thousandths of an inch. The use of precision gears is not in itself sufficient to eliminate backlash. Other factors must be considered. The following is a list of important factors which affect backlash:

1. Total composite error, TCE (refer to Sect. 7.18)
2. Tooth-to-tooth composite error, TTCE (refer to Sect. 7.18)
3. Center distance tolerance
4. Parallelism of gear axes
5. Types of bearings and subsequent wear
6. Side runout or wobble
7. Deflection under load
8. Gear tooth wear
9. Thermal expansion of gears and housing

Another method for reducing backlash is to use a spring-loaded antibacklash gear. This gear (Fig. 7.31) consists of two gears connected with springs; one gear is floating and the other is fixed to the hub. The pinion or mating gear nests between the spring-loaded gears. The spring-loaded gears must maintain a com-

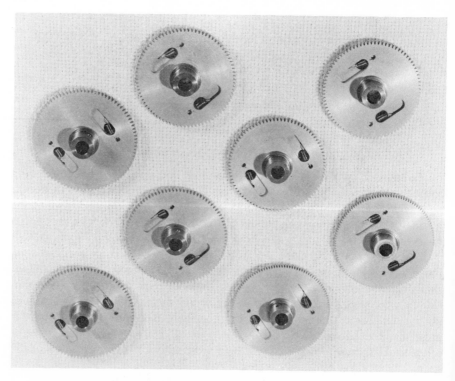

Fig. 7.31. Spring-loaded antibacklash gears. (Courtesy of Winfred M. Berg, Inc.)

pressive force on the pinion while in operation. Spring-loaded antibacklash gears are limited to low-torque applications.

In a reduction gear train (gear reducer), the last pair of gears, which is near the load, contributes the most toward the total backlash of the system. Therefore, the rule is to keep the backlash of the last pair of gears as small as possible. If appropriate, let the antibacklash gear be the last gear in the gear train.

7.17 Class and Quality Number

To assist the gear user and manufacturer, the AGMA has established a class number, Standard 390.02, which covers coarse- and fine-pitch spur, helical, herringbone, bevel, and hypoid gears. The class number consists of a quality number (identifying gear tolerances) and a material and treatment number. For example, a complete class number for a fine pitch gear may be 10C-A-4, where 10 is the quality number, C is the backlash designation, and A-4 is the material and treatment number, which includes hardness. Nonmetallic gears are not specifically covered in the AGMA standard.

The quality number for coarse pitch gears range from 3 to 15, and fine pitch gears range from 5 to 16. The higher the quality number, the more precise the gearing will be and the closer the tolerances. Gear costs increase as closer

Gearing application (coarse pitch)	Quality number or range
Agriculture Farm elevator	3 - 7
Construction Ditch digger	3 - 8

Machine tool industry
  Power drives: 0 - 800 fpm
                800 - 2000 fpm
                2000 - 4000 fpm
                Over 4000 fpm
	6 - 8
	8 - 10
	10 - 12
	12 and up

Mining and preparation Conveyor	5 - 7
Paper and pulp Envelope machines	6 - 8
Steel industry Electric furnace tilt	5 - 6

Fig. 7.32. Applications and quality numbers for coarse pitch spur, helical, and herringbone gears. Quality numbers are inclusive, from lowest to highest numbers.

Gearing application (fine pitch)	Quality number or range
Commercial meters Liquid, water, milk	7 - 9

Computing and accounting machines
        Computing
        Data processing
	10 - 11
	7 - 9

Electronic instrument control and guidance systems
        Aircraft instrument
        Pressure Transducer
	12
	12 - 14

Servo system component	9 - 11
Home appliances Timer	8 - 10
Small power tools Drills and saws	7 - 9

Fig. 7.33. Applications and quality numbers for fine pitch spur, helical, and herringbone gears. Quality numbers are inclusive, from lowest to highest numbers.

tolerances are specified. A few applications and corresponding quality numbers are listed in Figs. 7.32 and 7.33. A more complete listing can be found in the AGMA standard.

7.18 Composite Error

There are many procedures and pieces of equipment used to inspect gears. A relatively low-cost and fast method of determining gear quality is the composite check. This check is run on a variable center distance gear-rolling device. The results are considered to fall into two categories: the tooth-to-tooth composite error (TTCE) and the total composite error (TCE). The total composite error includes the tooth-to-tooth composite error.

A schematic setup of the gear-rolling fixture is shown in Fig. 7.34. Before testing the gear, the center distance between shafts is established. If the dial indicator is being used, the dial is set to zero and a specified weight or tension is placed onto the lever arm. As the tested gear is rotated through 360 deg or more the master gear moves slightly in and out of mesh. This movement can be amplified and continuously monitored on a chart. One inherent source of error is the master gear; although the master gear is very accurately manufactured, it is still subject to manufacturing errors.

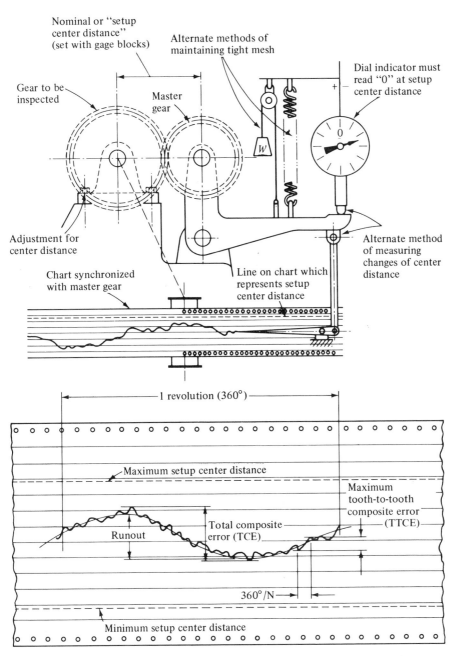

Fig. 7.34. Composite gear-checking setup and composite gear chart. (From Gear Handbook—the Design, Manufacture and Application of Gears by Darle W. Dudley. Courtesy of McGraw-Hill, Inc.)

275

The tooth-to-tooth composite error (refer to chart shown in Fig. 7.34) is a local error which is comprised of profile, tooth thickness, and lead errors. Profile error is the difference between the actual tooth form and the theoretical involute. Tooth thickness is an error in the tooth size. Lead error is the variation of the tooth shape along the width of the tooth.

The total composite error (refer to Fig. 7.34) is the difference between the highest and lowest reading; it consists of the pitch circle runout, lateral runout, and tooth-to-tooth composite error. Pitch circle runout is a measure of the pitch diameter eccentricity relative to the gear bore. Lateral runout can be described as the side wobble of a gear.

The AGMA quality number for fine pitch gears denotes the allowable TTCE and TCE. For example, a 1.75-in. 112-tooth fine pitch spur gear with a quality number of 10 must have a TTCE of less than .0005 in. and a TCE of less than .0010 in., according to AGMA 390.02.

7.19 Gear Manufacture

Cut gears are usually manufactured by hobbing or shaping. Both generating processes are suitable for high- and low-production runs. In a gear-generating process the gear teeth are formed by a series of cuts. An advantage of the generating process is that a hob or cutter of a given pitch can be used to cut spur and helical gears with any number of teeth. If an interference condition exists, the generating tool will undercut the gear teeth.

The hobbing process (Fig. 7.35) uses a "worm"-like cutter called a hob. The hob and workpiece rotate at a constant ratio as the hob is fed across the workpiece.

Fig. 7.35. Hobbing a spur gear. (Courtesy of The Fellows Gear Shaper Co.)

The shaping process (Fig. 7.36) uses a "pinion"-type cutter. As the cutter reciprocates, the cutter and the workpiece are intermittently indexed.

The simultaneous application of hobbing and shaping is called shobbing. Figure 7.37 shows the shobbing of a cluster spur gear.

Other gear-manufacturing methods are milling, broaching, punching, die casting, cold drawing, extruding, roll forming, powder metallurgy, and plastic molding.

Fig. 7.36. Shaping a helical gear. Shaper cutter is on left side of photograph. (Courtesy of The Fellows Gear Shaper Co.)

Fig. 7.37. Shobbing spur gears. Hob on left side, shaper cutters on right side. (Courtesy of The Fellows Gear Shaper Co.)

PROBLEMS

7.1 The speed ratio for a pair of spur gears is 2.5:1. The center distance is 1.40 in. Determine the pitch diameters.

7.2 The gear ratio for a pair of spur gears is 6:1. The pinion has 30 teeth. The center distance is $\frac{7}{8}$ in. Determine (a) the number of teeth on the gear, (b) the pitch diameters, (c) the diametral pitch, (d) the circular pitch.

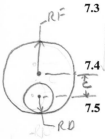

7.3 Determine the pitch diameters for a pair of standard 25-deg full depth spur gears. The driver rotates at 300 rpm and the driven gear at 100 rpm. The center distance is 2.10 in.

7.4 A 36-tooth spur gear drives a 96-tooth internal gear. The pitch is 12. Calculate the center distance between gear shafts.

7.5 In Fig. P7.5, a pivot arm moves a sheet of paper 180 deg from position 1 to position 2. The arm then returns to position 1. Four-bar linkage $ABCD$ drives spur gear $\textcircled{D}$. Gear $\textcircled{D}$ drives gear $\textcircled{F}$; the pivot arm is fastened to $\textcircled{F}$. Crank AB rotates ccw at a constant rpm. The gear ratio is 2.5:1. Determine (a) the linkage dimensions, (b) the pitch diameters, (c) the teeth on both gears (assume that the pitch is between 20 and 30).

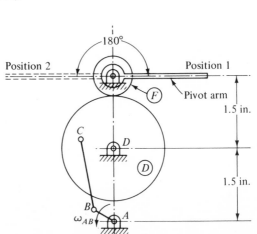

Fig. P7.5.

7.6 Construct three involute teeth for a full depth spur gear where the pressure angle is 20 deg on one side of the tooth profile and 25 deg on the other side. The pitch diameter is 8 in., and the diametral pitch is 3. Use the tooth proportions shown in Fig. 7.10.

7.7 A standard $14\frac{1}{2}$-deg full depth spur gear is driven by a rack. The spur gear has 16 teeth and a pitch of 16. Lay out the gears and determine (a) the maximum useful length of action and (b) the contact ratio based on part a.

7.8 The driver gear for a pair of standard 20-deg full depth spur gears has 20 teeth. The follower has 50 teeth, and the diametral pitch is 5. Lay out the gears and determine the contact ratio.

7.9 The length of action for a nonundercutting spur gear pair can be calculated from the equation $\sqrt{R_{o_1}^2 - R_{b_1}^2} + \sqrt{R_{o_2}^2 - R_{b_2}^2} - (R_1 + R_2) \sin \phi$, where the subscripts 1 and 2 represent gears 1 and 2. Calculate the contact ratio for the gear pair in Prob. 7.8.

7.10 A standard 20-deg full depth 40-tooth spur gear meshes with a 120-tooth internal gear. The diametral pitch is 48. Calculate the contact ratio using the equation

$$\frac{\sqrt{R_{o_1}^2 - R_{b_1}^2} - \sqrt{R_{i_2}^2 - R_{b_2}^2} + (R_2 - R_1) \sin \phi}{p \cos \phi}$$

where the subscript 2 refers to the internal gear. Refer to the symbols in Fig. P7.10; note that the internal radius $R_{i_2} = R_2 -$ addendum. Would the center distance tolerance and gear tolerances affect the contact ratio?

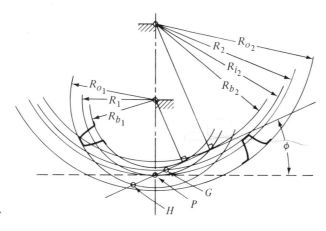

Fig. P7.10.

7.11 Figure P7.11 shows a portion of a hand-operated rolling press where gears $\textcircled{A}$ and $\textcircled{B}$ drive rollers $\textcircled{C}$ and $\textcircled{D}$. Both spur gears are standard 20-deg full depth gears having 19 teeth with a diametral pitch of 8. Gear pitch diameters and roller diameters are all the same dimension. The spacing between rollers can be adjusted by moving the "shaft support" vertically. Note that this adjustment changes the center distance between gears. Determine the maximum separation between rollers when the contact ratio is limited to 1.

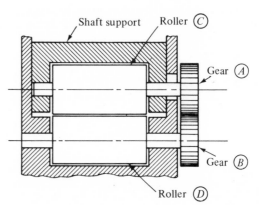

Fig. P7.11.

7.12 Select a spur gear pair from a gear catalog which meets the following conditions. The gear ratio is 2:1. The center distance is 3 in. The diametral pitch can be 6, 8, or 10. Give details concerning the teeth, diameter, material, hub projection, face width, and pressure angle of each gear.

7.13 A pair of standard $14\frac{1}{2}$-deg full depth spur gears are to have a gear ratio of 3:1. Assume that the diametral pitch is 16. Avoid interference. Determine (a) the number of teeth on each gear, (b) the diameter of each gear, (c) the center distance. Can these gears be found in a gear catalog?

7.14 The driver gear for a pair of standard 20-deg full depth spur gears rotates at 1800 rpm. The driven gear is to operate at approximately 800 rpm. The center distance is approximately 3.50 in. The diametral pitch range is 10 to 14. Determine a gear set which comes closest to an output of 800 rpm.

7.15 The driver gear for a pair of standard 20-deg full depth spur gears rotates at 1800 rpm. The driven gear is to operate at approximately 800 rpm. The center distance is approximately 1.00 in. The diametral pitch range is 50 to 60. Determine a gear set which comes closest to an output of 800 rpm.

7.16 The driver pinion for a pair of standard 25-deg full depth spur gears rotates at 1000 rpm. The larger gear is to rotate at approximately 325 rpm. The center distance is approximately 2.35 in. The diametral pitch is 14, 16, or 18. Determine a gear set which comes closest to an output of 325 rpm.

7.17 Design a spur gear pair to fit into a rectangular space approximately 5×2 in. The speed ratio is to be close to 5:14. The diametral pitch is limited to 24, 32, or 48.

7.18 Figure P7.18 shows a belt-driven spur gear pair. The pulley diameter is 2 in., and the belt speed is 1000 ft/min. Gear A is keyed to the pulley. The speed ratio for the gears is to be 2:1, and the diametral pitch is 9. The center distance is approximately 1.75 in. Determine (a) the rpm for gear B, (b) the number of teeth on each gear, (c) the pitch diameters, (d) the center distance.

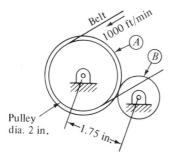

Fig. P7.18.

7.19 Figure P7.19 shows the drill and feed mechanisms of an electric drill used to drill structural steel. The drill mechanism consists of a flexible shaft, connected to an electric motor, which drives a bevel-spur gear train. The feed mechanism is driven by a hand-operated ratchet, driving through a pair of spur gears to a feed screw which carries the gear box. The gear to pinion tooth ratios are to be 1.5:1. The idler to pinion tooth ratio is 1:1. Assuming that the spur gears have a diametral pitch of 16, determine the pitch diameters and teeth for all spur gears.

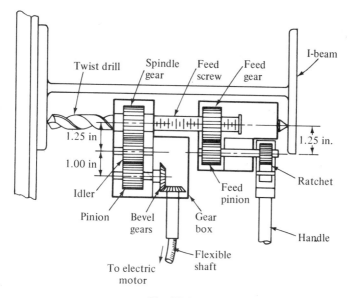

Fig. P7.19.

282 GEARS Chap. 7

7.20 An 18-tooth pinion drives a 24-tooth spur gear. The diametral pitch is 6. The pinion rotates at 1800 rpm. What is the probable torque of the driven gear and the driver gear? The horsepower input is 5. Use hp $= Tn/63,000$, where hp = horsepower, T = torque (in.-lb), and n = rpm.

7.21 A 50-tooth helical gear has a normal diametral pitch of 16 and a helix angle of 15 deg. Determine (a) the pitch diameter, (b) the transverse diametral pitch, (c) the transverse circular pitch.

7.22 A parallel helical gear pair has a helix angle of 14°32′. The center distance is 3.099 in. One gear has 20 teeth; the other gear 40 teeth. Determine the normal and transverse diametral pitches.

7.23 Design a helical gear pair to connect parallel shafts which are 2.000 in. apart. The gear ratio is to be 7:1. The normal diametral pitch ranges from 20 to 30. Use a normal pressure angle of 25 deg and assume the pinion to have 11 teeth.

7.24 Design a helical gear pair to connect parallel shafts which are 2.000 in. apart. The driver rotates at 5000 rpm, and the follower at 1825 rpm. The normal diametral pitch ranges between 48 to 72. Also, calculate the pitch-line velocity.

7.25 Select two helical gears from a gear catalog to connect parallel shafts 3 in. apart. The driver receives 3 hp at 900 rpm. The speed ratio is 1:2. The follower shaft can be supported only on the right side (Fig. P7.25). Give details concerning pitch diameters, normal and transverse pitches, helix angle, pressure angle, hand of helix, and materials.

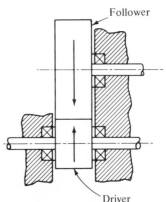

Fig. P7.25.

7.26 It is desirable to connect two parallel shafts with a helical gear pair. The center distance is 3.000 in., and the gear ratio is 10:1. Determine a reasonable minimum value for the normal diametral pitch (largest tooth size). State all assumptions.

7.27 Determine the hand and thrust directions (from gear to shaft) for each helical gear shown in Fig. P7.27. The direction of rotations are indicated in the figure.

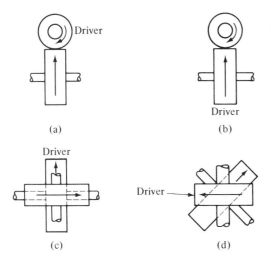

Fig. P7.27. (a) (b) (c) (d)

7.28 The shaft angle for a pair of crossed helical gears is 90 deg. Both gears have a 45-deg helix angle. The normal diametral pitch is 10. The gear ratio is 2:1, and the driver (pinion) has 15 teeth. Determine (a) the center distance, (b) the teeth on the follower gear, (c) the pitch diameters.

7.29 Determine a pair of crossed helical gears to connect shafts at $\Sigma = 90$ deg. The speed ratio is 2.5:1. The center distance is approximately $1\frac{3}{4}$ in., and the normal diametral pitch is 16.

7.30 Use the same information as Prob. 7.29 except that the center distance is to be exactly 1.750 in. Determine the gear pair.

7.31 A crossed helical gear pair is to connect two shafts at a shaft angle of 60 deg. The gear ratio is 2.4:1. The center distance is approximately 2 in. The normal diametral pitch is 20. Determine the gear pair.

7.32 A pair of crossed helical gears are to be mounted at a shaft angle of 90 deg. The gear ratio is 3:1. The normal diametral pitch is to be 48 or 64. Select gears from a gear catalog. From the rotations shown in Fig. P7.32, determine the best positions for thrust washers.

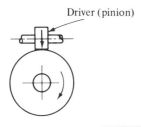

Fig. P7.32.

7.33 A four-thread, right-hand worm drives a 40-tooth and 50-tooth worm gear (Fig. P7.33). The worm rotates at 1000 rpm. Determine (a) the rpm of each worm gear and (b) the direction of rotation for each gear.

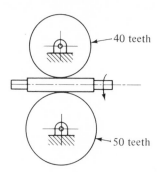

40 teeth

50 teeth

Fig. P7.33.

7.34 Determine the peripheral speed (in./min) for one of the grinding wheels shown in Fig. P7.34. The electric motor is rated at 1800 rpm. Both worms have four threads. Each worm gear has 20 teeth. The outside diameter of the grinding wheel is 1.50 in.

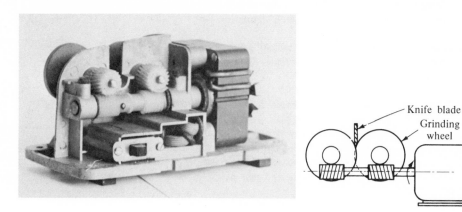

Knife blade

Grinding wheel

Fig. P7.34.

7.35 Figure P7.35 shows a simplified sketch of a washing machine agitator. Oscillating motion is obtained from the scotch yoke mechanism. An electric motor drives pulley Ⓐ at 1800 rpm. Pulley diameter Ⓐ is 3 in., and Ⓑ is 5 in. The worm is triple thread and the worm gear contains 57 teeth. Determine the number of oscillations per minute from the agitator shaft.

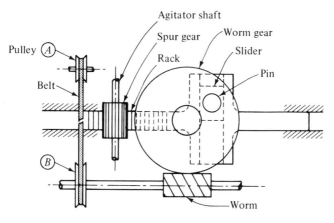

Fig. P7.35.

7.36 The speed ratio for a worm gearset is to be 1:50. The driver is the worm, and it rotates at 1000 rpm. The center distance is approximately .625 in. Select gears from a catalog; use a pitch of 48. Determine (a) the exact center distance, (b) the lead angle of the worm, (c) the pitch-line velocity for the worm and worm gear.

7.37 Select a pair of straight miter gears from a gear catalog. The pitch is 16, and the mounting distance is approximately $1\frac{3}{8}$ in. Determine (a) the pitch diameters, (b) the actual mounting distance, (c) the number of teeth on each gear.

7.38 Figure P7.38 shows a bevel gear differential. Each bevel gear has 80 teeth. The T-shaped arm receives 10 rpm. Bevel gear C is stationary. Note that the gears are not keyed or fastened to the arm. Determine the rpm for gear A.

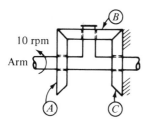

Fig. P7.38.

7.39 A relationship between backlash and change in center distance for a pair of spur gears is $B = 2(\Delta C)\tan\phi$. The angular backlash is $\beta = B(360°)/\pi D$. The symbols are B = backlash or linear backlash measured on the pitch circle (in.), β = angular backlash (deg), ΔC = change in center distance (in.), ϕ = pressure angle (deg), and D = pitch diameter of gear or pinion (in.). Determine the linear and angular backlash for a pair of 1-in. spur gears. The difference between the standard (theoretical) and actual center distance is .002 in., and the pressure angle is 20 deg.

7.40 The normal backlash B_n, measured along the line of action, between a pair of coarse 20-deg full depth spur gears is .001 in. (see Fig. P7.40). Determine (a) the pitch circle backlash B using the equation $B = B_n/\cos\phi$ and (b) the angular backlash for a 2-in. pitch diameter gear (refer to equation shown in Prob. 7.39).

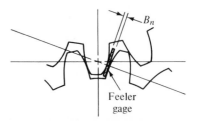

Feeler gage

Fig. P7.40.

7.41 Determine the allowable tooth-to-tooth composite error and the total composite error for a 100-tooth spur gear which is to be used in a servo system. The pitch diameter is slightly over 2 in. The solution requires the use of AGMA 390.02 or a suitable handbook. *Note*: Select a reasonable quality number.

Chapter **8**

GEAR TRAINS

8.1 Introduction

A drive consisting of gears is commonly called a gear train. There are two broad categories of gear trains, ordinary and planetary. In an ordinary gear train all the gear axes are fixed relative to the frame (Figs. 8.1 and 8.3). In a planetary gear train one or more of the gear axes rotate relative to the frame (Fig. 8.14).

8.2 Simple Gear Train

The ordinary gear train can be either simple or compound. In the simple type, each shaft carries one gear (Fig. 8.1). In the compound type, discussed in the next section, one or more of the shafts carry two gears.

Figure 8.1 shows a simple gear train driven by gear $\textcircled{A}$. The speed ratio for successive mating pairs is

$$\frac{\omega_B}{\omega_A} = \frac{N_A}{N_B}, \qquad \frac{\omega_C}{\omega_B} = \frac{N_B}{N_C}, \qquad \frac{\omega_D}{\omega_C} = \frac{N_C}{N_D}, \qquad \frac{\omega_E}{\omega_D} = \frac{N_D}{N_E}$$

287

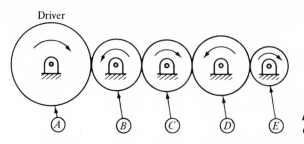

Driver

Fig. 8.1. *Simple gear train. Gear Ⓐ is the driver.*

The overall angular velocity ratio (ω_E/ω_A) can be written as

$$\frac{\omega_E}{\omega_A} = \left(\frac{\omega_B}{\omega_A}\right)\left(\frac{\omega_C}{\omega_B}\right)\left(\frac{\omega_D}{\omega_C}\right)\left(\frac{\omega_E}{\omega_D}\right) \tag{1}$$

Note that ω_B, ω_C, and ω_D divide out of Eq. (1). Equation (1) is not too useful in its present form; therefore the tooth ratios are substituted into the right-hand side of Eq. (1). Now,

$$\frac{\omega_E}{\omega_A} = \left(\frac{N_A}{N_B}\right)\left(\frac{N_B}{N_C}\right)\left(\frac{N_C}{N_D}\right)\left(\frac{N_D}{N_E}\right)$$

or, simply,

$$\frac{\omega_E}{\omega_A} = \frac{N_A}{N_E} \tag{2}$$

Gears Ⓑ, Ⓒ, and Ⓓ do not appear in Eq. (2) and thus do not contribute to the overall angular velocity ratio. These gears are termed idler gears.

A simple gear train may be employed to fill in space between two gears, or to reverse the direction of rotation for the last driven gear, or in a multiple drive application, the intermediate gears are used as power takeoffs.

EXAMPLE 8.1

Given: Simple gear train (Fig. 8.2), where gear Ⓐ rotates 1000 rpm cw.

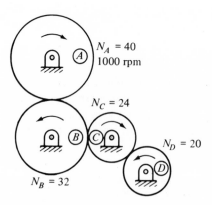

$N_A = 40$
1000 rpm

$N_C = 24$

$N_D = 20$

$N_B = 32$

Fig. 8.2. *Simple gear train.*

Rpm is angler Velocity

Determine: Rpm of gear ⓓ.

Solution

$$\frac{\omega_D}{\omega_A} = \frac{N_A}{N_D}$$

$$\omega_D = \omega_A \left(\frac{N_A}{N_D} \right) = 1000 \left(\frac{40}{20} \right) = 2000 \text{ rpm ccw}$$

8.3 Compound Gear Train

A compound gear train is shown in Fig. 8.3, note that the gears are fastened together on the same shaft. The speed ratio for successive mating pairs is

$$\frac{\omega_B}{\omega_A} = \frac{N_A}{N_B}, \qquad \frac{\omega_D}{\omega_C} = \frac{N_C}{N_D}, \qquad \frac{\omega_F}{\omega_E} = \frac{N_E}{N_F}$$

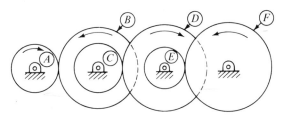

Fig. 8.3. *Compound gear train. Gear ⓐ is the driver.*

The overall angular velocity ratio can be written as

$$\frac{\omega_F}{\omega_A} = \left(\frac{\omega_B}{\omega_A} \right) \left(\frac{\omega_D}{\omega_C} \right) \left(\frac{\omega_F}{\omega_E} \right) \tag{1}$$

At first glance, Eq. (1) seems to show that the left-hand side is not equal to the right-hand side; however, consider that ω_B, ω_C, ω_D, and ω_E divide out of the equation because

$$\omega_B = \omega_C \qquad \text{and} \qquad \omega_D = \omega_E$$

Substituting tooth ratios into the right-hand side of Eq. (1), we obtain

$$\frac{\omega_F}{\omega_A} = \left(\frac{N_A}{N_B} \right) \left(\frac{N_C}{N_D} \right) \left(\frac{N_E}{N_F} \right) \tag{2}$$

In this equation all the gears enter into the overall angular velocity ratio.

There is one angular velocity equation which can be used either for simple or compound gear trains. Recall that the speed ratio for a pair of gears was defined as

$$\frac{\omega_F}{\omega_D} = \frac{N_D}{N_F}$$

In a similar fashion the speed ratio for an originary gear train can be written as

$$\frac{\omega_{fo}}{\omega_{dr}} = \frac{\times N_{dr}}{\times N_{fo}} \tag{8.1}$$

Where ω_{fo} and ω_{dr} denote the angular velocity of the follower (last driven gear in the gear train) and the driver (first driver in the gear train). The tooth ratio, symbolized as $\times N_{dr}/\times N_{fo}$, represents the product of all the driver teeth divided by the product of all the follower teeth. To be considered a driver-follower pair, the gear teeth must mesh. Equation (8.1) can be extended to chain and belt drives. For flat belt drives, use pulley diameters; for V-belt drives, use pitch diameters.

To illustrate the use of Eq. (8.1), consider the compound gear train in Fig. 8.3. Gear Ⓕ is the gear train follower, and gear Ⓐ is the driver; thus $\omega_{fo} = \omega_F$ and $\omega_{dr} = \omega_A$. The individual driver-follower gears which constitute the tooth ratio are determined as follows. Gear Ⓐ is in mesh with gear Ⓑ, where gear Ⓐ drives Ⓑ; thus Ⓐ is the driver and Ⓑ is the follower. Gear Ⓒ is in mesh with Ⓓ; Ⓒ is a driver and Ⓓ is a follower. Last, Ⓔ is a driver and Ⓕ is a follower. Using Eq. (8.1),

$$\frac{\omega_{fo}}{\omega_{dr}} = \frac{\times N_{dr}}{\times N_{fo}}$$

Substitute into this equation, we obtain

$$\frac{\omega_F}{\omega_A} = \left(\frac{N_A}{N_B}\right)\left(\frac{N_C}{N_D}\right)\left(\frac{N_E}{N_F}\right) \tag{3}$$

Equation (3) is exactly the same as Eq. (2).

EXAMPLE 8.2

Given: Figure 8.4, is a quadruple reduction helical gear train driven by an electric motor rated as 1750 rpm. Gear teeth are $N_A = 26$, $N_B = 70$, $N_C = 20$, $N_D = 100$, $N_E = 16$, $N_F = 104$, $N_G = 20$, and $N_H = 80$.

Determine: Output shaft rpm.

Solution. A look at Fig. 8.4 shows that the last gear pair (low-rpm end) has coarse teeth and a wider face width than the first gear pair (1750-rpm end). Although all the gears in the gear train transmit approximately the same horsepower, the low-rpm gears transmit higher torque. The high torque and accompanying high gear forces require a stronger gear.

Use Eq. (8.1), where the gear train follower is gear Ⓗ and the driver is Ⓐ. The individual driver-follower pairs required for the tooth ratio can be easily identified in Fig. 8.4. For example, the first tooth ratio, starting from driver Ⓐ, is N_A/N_B. The complete solution for this example is

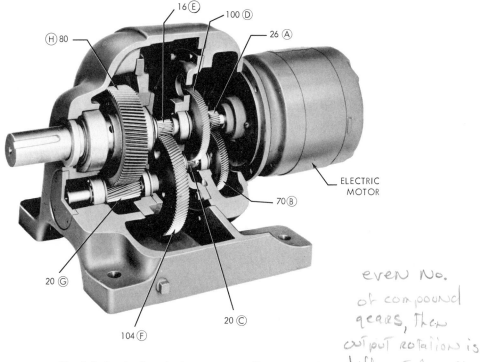

Fig. 8.4. *Quadruple reduction gearmotor.* (Courtesy of Link-Belt.)

[handwritten note:] even No. of compound gears, Then output rotation is different than the input rotation

$$\omega_{fo} = \omega_{dr} \frac{\times N_{dr}}{\times N_{fo}}$$

$$\omega_H = \omega_A \left(\frac{N_A}{N_B}\right)\left(\frac{N_C}{N_D}\right)\left(\frac{N_E}{N_F}\right)\left(\frac{N_G}{N_H}\right)$$

$$\omega_H = 1750 \left(\frac{26}{70}\right)\left(\frac{20}{100}\right)\left(\frac{16}{104}\right)\left(\frac{20}{80}\right) = 5 \text{ rpm}$$

The output shaft rotates in the same direction as the input.

EXAMPLE 8.3

Given: Figure 8.5 shows the table-rotating mechanism of a welding positioner. A variable-speed motor drives sprocket Ⓐ at 800 rpm.

Determine: The time required for the table to rotate through $\frac{1}{2}$ rev when sprocket Ⓐ rotates at 800 rpm.

Solution. We shall not consider the time required to accelerate or decelerate the table owing to inertia.

Equation (8.1) can be used to solve this problem:

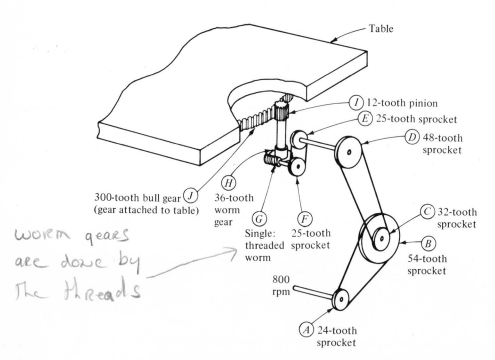

Table

I 12-tooth pinion
E 25-tooth sprocket
D 48-tooth sprocket

300-tooth bull gear J
(gear attached to table)

H 36-tooth worm gear

C 32-tooth sprocket

G Single: threaded worm

F 25-tooth sprocket

B 54-tooth sprocket

800 rpm

A 24-tooth sprocket

worm gears are done by the threads

Fig. 8.5. *Table-rotating mechanism of a heavy-duty (175-ton capacity) welding positioner.* (Courtesy of Aronson Machine Company, Inc.)

292

$$\omega_{fo} = \omega_{dr} \frac{\times N_{dr}}{\times N_{fo}}$$

$$\omega_J = \omega_A \left(\frac{N_A}{N_B}\right)\left(\frac{N_C}{N_D}\right)\left(\frac{N_E}{N_F}\right)\left(\frac{thd}{N_H}\right)\left(\frac{N_I}{N_J}\right)$$

$$\omega_J = 800 \left(\frac{24}{54}\right)\left(\frac{32}{48}\right)\left(\frac{25}{25}\right)\left(\frac{1}{36}\right)\left(\frac{12}{300}\right)$$

$$\omega_J = \frac{1600}{6075} \text{ rpm}$$

To determine the time for $\frac{1}{2}$ rev of $\textcircled{J}$, use the equation

$$t = \frac{\theta}{\omega} = \frac{.5 \text{ rev}}{(1600 \text{ rev}/6075 \text{ min})} = 1.9 \text{ min}$$

8.4 Speed Changers—Ordinary Gear Train Type

There are various kinds of speed changers or transmissions, as they are usually called. These can be categorized as gear, belt, chain, linkage, hydraulic, and electrical, or a combination of these. A transmission either can be designed into a machine or mechanism or a packaged-type transmission can be attached to the machine. Our concern in this section and the following section is with ordinary gear train transmissions. Characteristically, an ordinary geared transmission can transmit high power at high efficiency with no slip.

An example of a geared transmission is shown in Fig. 8.6. This is a sliding gear transmission, where sliding gears $\textcircled{C}$ and $\textcircled{D}$ can slide back and forth along the splined countershaft. Rotational motion can be directly transmitted from the splined shaft to the internal spline of the sliding gears. The input shaft is connected to gear $\textcircled{A}$, and $\textcircled{A}$ and $\textcircled{B}$ are always in mesh. Gears $\textcircled{E}$ and $\textcircled{F}$ are rigidly connected to the output shaft. The left end of the output shaft turns freely inside the input shaft. There are three positions for the sliding gears: neutral (zero

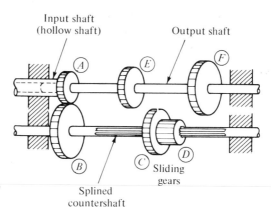

Fig. 8.6. Neutral position for sliding gear transmission.

output speed), low, and high; Fig. 8.6 shows the neutral position. A low output speed is obtained by shifting sliding gear D into mesh with F. High speed is obtained when C meshes with E. To reduce tooth clash when shifting from low to high speed or vice versa, the input shaft is either disengaged from the source of power, by means of a clutch, or the power source is turned off. A clutch is a device which can connect and disconnect power. The gear train speed ratio for low speed is

$$\frac{\omega_{fo}}{\omega_{dr}} = \left(\frac{N_A}{N_B}\right)\left(\frac{N_D}{N_F}\right)$$

For high speed,

$$\frac{\omega_{fo}}{\omega_{dr}} = \left(\frac{N_A}{N_B}\right)\left(\frac{N_C}{N_E}\right)$$

Another type of geared transmission is the constant mesh transmission (Fig. 8.7). The particular transmission shown employs electromagnetic clutches.

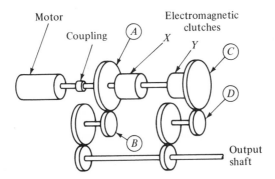

Fig. 8.7. Constant mesh transmission.

When electromagnetic clutch X is energized and clutch Y is de-energized, gear A drives gear B. When clutch Y is energized and X de-energized, C drives D. If both clutches are de-energized, no power is transmitted to the output shaft. This system eliminates tooth clashing because the gears are always in mesh.

There are many other types of geared transmissions. The manually shifted automotive transmission is discussed in the next section.

EXAMPLE 8.4

Given: Figure 8.8 shows a portion of a gear drive mechanism used in a toy truck.

Determine: The angular velocity of the electric motor with respect to the angular velocity of one of the wheels. Calculate this ratio for low and high speeds.

Solution. This mechanism provides two speeds, both in forward and reverse. Gears B, C, F, and H rotate freely on a squared shaft. Gear I is rigidly fastened to this shaft. Speed can be changed by means of a shift lever. The shift lever controls a double-

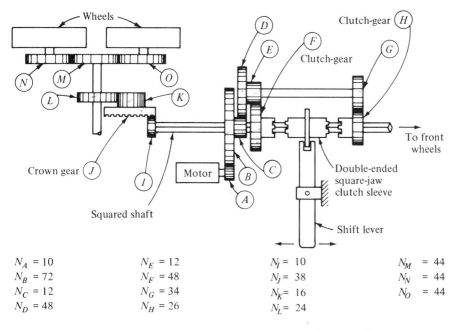

N_A = 10	N_E = 12	N_I = 10	N_M = 44
N_B = 72	N_F = 48	N_J = 38	N_N = 44
N_C = 12	N_G = 34	N_K = 16	N_O = 44
N_D = 48	N_H = 26	N_L = 24	

Fig. 8.8. Gear drive mechanism of a toy truck. (Courtesy of Remco Industries, Inc.)

ended square-jaw clutch sleeve. The clutch sleeve has a square bore and can slide on the squared shaft. When the sleeve is engaged with either clutch gear Ⓕ or Ⓗ, power can be transmitted along the squared shaft to gear Ⓘ. Shifting can take place with the gears rotating. The gear train is reversed by reversing the polarity of the dc motor.

Low speed is accomplished by engaging clutch gear Ⓕ. The reciprocal of the low speed ratio is

$$\frac{\omega_A}{\omega_N} = \frac{\times N_{\text{fo}}}{\times N_{\text{dr}}} = \left(\frac{72}{10}\right)\left(\frac{48}{12}\right)\left(\frac{48}{12}\right)\left(\frac{38}{10}\right)\left(\frac{24}{16}\right)\left(\frac{44}{44}\right) = 656.64$$

For high speed, engage clutch gear Ⓗ. The reciprocal of the high speed ratio is

$$\frac{\omega_A}{\omega_N} = \frac{\times N_{\text{fo}}}{\times N_{\text{dr}}} = \left(\frac{72}{10}\right)\left(\frac{48}{12}\right)\left(\frac{26}{34}\right)\left(\frac{38}{10}\right)\left(\frac{24}{16}\right)\left(\frac{44}{44}\right) = 125.53$$

8.5 Manually Shifted Automotive Transmissions

The conventional, manually shifted automobile transmission is a three-speed forward and one-speed reverse unit. Forward speed positions are referred to as low, second (intermediate), and high. Low speed is used to start the car moving smoothly from rest. Here the engine provides the wheels with high torque, which is required to overcome the car's deadweight and frictional resistance.

Once the car is in motion, the operator shifts into second and then into high for high speed.

Gear clashing is practically eliminated in the modern manually shifted automobile transmission by the use of a clutching device called a synchronizer. A synchronizer, as the name implies, synchronizes the speed of a gear and shaft just prior to the transmission of power from the engine. Although there are many synchronizers presently being used, only one type will be discussed here.

Figure 8.9 shows a plain-type synchronizer. This figure is part of the three-speed transmission shown in Fig. 8.10. The hub is splined internally to the main shaft and externally to the sliding sleeve. Spring-loaded balls positioned radially around the hub retain the sliding sleeve to the hub. The splined helical gear rotates freely on the main shaft.

The following discussion concerns the operation of a plain-type synchronizer in a transmission [refer to Fig. 8.9(a)]. The operator disengages the engine clutch; this disconnects the engine from the transmission and then manually moves the shift fork to the right. This moves the sliding sleeve, along with the hub, to the right until the conical surfaces on the hub and helical gear make contact. Friction between these surfaces quickly causes the gear and hub to rotate as one unit. Further motion to the right overcomes the spring forces, which allows the sliding sleeve to move partly off of the hub and into engagement with the external splines of the helical gear [Fig. 8.9(b)]. Now the engine clutch is engaged and

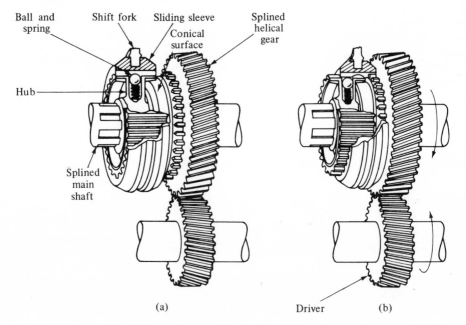

Fig. 8.9. Plain-type synchronizer. (a) Prior to synchronization; (b) when fully engaged.

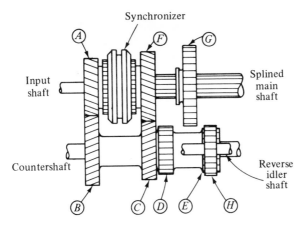

Fig. 8.10. *Three-speed transmission, synchronized in second and high.*

power is transmitted through the helical gears to the sliding sleeve and hub and then onto the main shaft. Note that the ends of the splines are rounded so that the gear and sliding sleeve can be aligned and then interlocked.

Figure 8.10 shows a manually shifted three-speed transmission which uses a synchronizer for second and high speed. This transmission is a combination of the sliding gear and constant mesh types. Before shifting into gear, the engine clutch temporarily disengages the engine from the transmission's input shaft. The input shaft is part of gear Ⓐ. Gear Ⓕ spins freely on a bearing which is carried by the main shaft. All the countershaft gears rotate together.

Low. Low is obtained by moving sliding gear Ⓖ into mesh with Ⓓ. The gear train speed ratio for low is

$$\frac{\omega_G}{\omega_A} = \left(\frac{N_A}{N_B}\right)\left(\frac{N_D}{N_G}\right)$$

Second. For second, the synchronizer is moved toward Ⓕ, synchronizing it and locking it to the main shaft. The gear train speed ratio for second is

$$\frac{\omega_F}{\omega_A} = \left(\frac{N_A}{N_B}\right)\left(\frac{N_C}{N_F}\right)$$

High. For high, the synchronizer is moved toward Ⓐ, synchronizing it and locking it to the main shaft. This is direct drive; the main shaft rotates at the same rpm as the input shaft.

Reverse. For reverse, sliding gear Ⓖ is meshed with idler gear Ⓗ. Note that gears Ⓔ and Ⓗ are always in mesh. The gear train speed ratio for reverse is

$$\frac{\omega_G}{\omega_A} = \left(\frac{N_A}{N_B}\right)\left(\frac{N_E}{N_H}\right)\left(\frac{N_H}{N_G}\right) = \left(\frac{N_A}{N_B}\right)\left(\frac{N_E}{N_G}\right)$$

Cancellation of N_H indicates that Ⓗ is an idler gear. It should be realized that the differential, which is connected between the transmission and the wheels, will further alter the engine speed.

Many of the present-day manually shifted automobile transmissions are fully synchronized in all forward speeds. These transmissions are available in either three or four forward speeds.

8.6 Design of a Compound Gear Train

So far this chapter has dealt mainly with the analysis of gear trains; given a gear train, determine the speed ratio. The usual type of problem encountered is the reverse of this; given a speed ratio, determine the gear train.

There are a number of methods for finding the number of gears and teeth in a compound gear train. One of the simpler methods will be presented here. This method, as well as most of the others, does not directly utilize a diametral pitch (or pitches) or a center distance (or distances).

To illustrate this method, let us solve the following problem. An instrument gear train requires a speed ratio of $1:17$. It has been decided to use standard 20-deg full depth spur gears with a diametral pitch of 32. Gears are to have no fewer than 16 teeth and no more than 96 teeth. Although the 16-tooth gear is slightly undercut, it is considered to be insignificant. The 96-tooth gear was chosen on the basis that a 3-in. pitch diameter ($D = N/P = 96/32 = 3$ in.) is a rather large gear for the application intended.

Using 16 teeth and 96 teeth, the maximum ratio for a single gear pair is $16/96 = 1/6$. A double pair is $(16/96)(16/96) = 1/36$. Since the desired ratio $1/17$ falls between $1/6$ and $1/36$, at least two gear pairs are needed.

To start with, the gear ratios are made equal:

$$\frac{\omega_{fo}}{\omega_{dr}} = \frac{1}{4.12}\left(\frac{1}{4.12}\right) \approx \frac{1}{17} \tag{1}$$

where $1/4.12$ was obtained from

$$\sqrt[2]{\frac{1}{17}} \approx \frac{1}{4.12}$$

A trial-and-error procedure now begins. Change one of the ratios in Eq. (1) to a number which is close to $1/4.12$, with the restriction that this number can be converted into a simple fraction. Try $1/4.125$. Note that the other ratio in Eq. (1) must be altered to maintain a speed ratio of $1/17$. The speed ratio equation now appears as

$$\frac{1}{4.125}\left(\frac{4.125}{17}\right) = \frac{1}{17} \qquad \text{or} \qquad \frac{1}{\frac{33}{8}}\left(\frac{\frac{33}{8}}{17}\right) = \frac{1}{17}$$

Simplify to

$$\frac{8}{33}\left[\frac{33}{8(17)}\right] = \frac{1}{17}$$

The actual tooth numbers are obtained by multiplying the numerator and denominator by 2; this results in

$$\frac{16}{66}\left(\frac{33}{136}\right) = \frac{1}{17}$$

This solution, of course, is not acceptable because 136 teeth is greater than 96. Let us try 1/4; now

$$\frac{1}{4}\left(\frac{4}{17}\right) = \frac{1}{17}$$

Multiplying the numerator and denominator by 16 and 4, we obtain

$$\frac{16}{64}\left(\frac{16}{68}\right) = \frac{1}{17}$$

This solution is acceptable. Thus the gear train consists of two 16-tooth gears, a 64-tooth gear, and a 68-tooth gear. The gear arrangement is shown in Fig. 8.11. Other satisfactory solutions are possible. The results obtained from this method can be applied to spur, helical, and bevel gears.

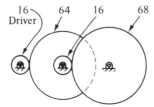

Fig. 8.11. *Compound gear train with a speed ratio of 1 : 17.*

If the method just presented does not lead to a satisfactory solution, the reader may want to refer to a broader method. Such a method can be found in Buckingham, "Gear Ratio Tables for 4-, 6-, and 8-Gear Combinations" (The Industrial Press, New York, 1958). This book contains a Brocot table which gives the decimal value (less than 1) for any combination of two gears each of which has 120 teeth or less. Here is a sample of a few values found in the table: $35/93 = .37634409$, $40/101 = .39603960$, $45/113 = .39823009$.

EXAMPLE 8.5

Given: The output shaft of a geared speed increaser is to rotate at 1500 rpm. Input shaft rotates at 200 rpm. No gear is to have less than 20 teeth or more than 50 teeth.

Determine: The number of gears and the number of teeth on each gear in this gear train.

Solution

STEP 1. The speed ratio for the geared increaser is

$$\frac{\omega_{fo}}{\omega_{dr}} = \frac{1500}{200} = 7.50$$

STEP 2. Determine the minimum number of gear pairs required. Using 20 and 50 teeth,

$$\text{One pair:} \quad \frac{50}{20} = 2.50$$

$$\text{Two pairs:} \quad \frac{50}{20}\left(\frac{50}{20}\right) = 6.25$$

$$\text{Three pairs:} \quad \frac{50}{20}\left(\frac{50}{20}\right)\left(\frac{50}{20}\right) = 15.63$$

At least three pairs are required because the desired speed ratio 7.50 is larger than 6.25 but smaller than 15.63.

STEP 3. To establish three equal pairs, take the cube root of 7.50:

$$\sqrt[3]{7.50} \approx 1.96$$

Thus, $1.96(1.96)(1.96) \approx 7.50$.

STEP 4. Change two of the 1.96 ratios to 2 and appropriately alter the third pair. The speed ratio equation now is

$$2(2)\left(\frac{7.50}{4}\right) = 7.50$$

STEP 5. To obtain tooth numbers, multiply the numerator and denominator by 20, 20, and 6:

$$\frac{40}{20}\left(\frac{40}{20}\right)\left(\frac{45}{24}\right) = 7.50$$

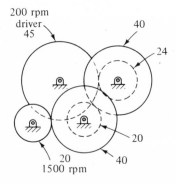

Fig. 8.12. Geared speed increaser.

This solution is satisfactory. We require two 20-tooth gears, two 40-tooth gears, a 24-tooth gear, and a 45-tooth gear. The gear train is laid out in Fig. 8.12.

8.7 Design of a Reverted Gear Train

A reverted gear train (Fig. 8.13) is a compound gear train where the input and output shafts are along the same line. Reverted trains are widely used because of their compactness and simplicity.

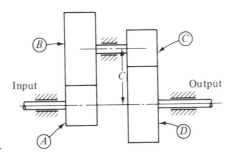

Fig. 8.13. Reverted gear train.

In the reverted gear train shown in Fig. 8.13, gears Ⓑ and Ⓒ are keyed to the same shaft. This shaft is parallel to the input-output shafts. The center distance, C, for both mating gear pairs is the same. The center distance in terms of pitch radii is

$$C = R_A + R_B = R_C + R_D$$

or

$$2C = D_A + D_B = D_C + D_D \tag{1}$$

Recall that the pitch diameter for a spur gear is $D = N/P$. Substitute for D in equation (1):

$$2C = \frac{N_A}{P_1} + \frac{N_B}{P_1} = \frac{N_C}{P_2} + \frac{N_D}{P_2} \tag{8.2}$$

where P_1 is the diametral pitch for mating gears Ⓐ and Ⓑ, and P_2 for gears Ⓒ and Ⓓ. Equation (8.2) can be simplified by assuming both gear pairs to have the same diametral pitch; then

$$2C = \frac{N_A}{P} + \frac{N_B}{P} = \frac{N_C}{P} + \frac{N_D}{P}$$

or

$$2CP = N_A + N_B = N_C + N_D \tag{8.3}$$

The implication of Eq. (8.3) is that in the speed ratio equation

$$\frac{\omega_{fo}}{\omega_{dr}} = \left(\frac{N_A}{N_B}\right)\left(\frac{N_C}{N_D}\right)$$

the sum of the numerator and denominator for each pair must be equal; that is, $N_A + N_B = N_C + N_D$.

Let us design a reverted spur gear train for a speed ratio of 1:18. The minimum number of teeth per gear is 16, and the maximum is 96. All gears are to have the same pitch. The center distance is not specified. The maximum ratio for one pair is $16/96 = 1/6$, and for two pairs $(16/96)(16/96) = 1/36$. Since $1/18$ falls between $1/6$ and $1/36$, use two pairs. The ideal value for each pair is $\sqrt[2]{1/18} \approx 1/4.25$. Avoid the use of large numbers in the ratios; the reason for this will become apparent later on. Let us start by using the following ratios:

$$\frac{1}{4}\left(\frac{4}{18}\right) = \frac{1}{18}$$

or

$$\frac{1}{4}\left(\frac{2}{9}\right) = \frac{1}{18}$$

The sum of the numerator and denominator for each ratio is 5 and 11. Obviously these values are not equal and thus will not conform to Eq. (8.3). To make these sums equal, multiply $1/4$ by $11/11$ and multiply $2/9$ by $5/5$, where 11 and 5 are obtained from the initial sum:

$$\left[\frac{1}{4}\left(\frac{11}{11}\right)\right]\left[\frac{2}{9}\left(\frac{5}{5}\right)\right] = \frac{1}{18}$$

$$\frac{11}{44}\left(\frac{10}{45}\right) = \frac{1}{18}$$

Now the sums are equal, $11 + 44 = 55$ and $10 + 45 = 55$. To obtain the actual tooth numbers and also to retain the sum, both the numerator and denominator of both ratios must be multiplied by the same number. Multiplying each by 2, we obtain

$$\frac{22}{88}\left(\frac{20}{90}\right) = \frac{1}{18}$$

The sum, $22 + 88 = 20 + 90 = 110$ teeth. This solution is acceptable. Referring to Fig. 8.13, gear Ⓐ has 22 teeth, Ⓑ has 88, Ⓒ has 20, and Ⓓ has 90.

8.8 Planetary Equation

Figure 8.14 shows a simple planetary gear train. The identifying feature of most planetary trains is the rotating arm or planet carrier. In this particular planetary, the sun gear is assumed stationary. As the arm rotates cw about O, the planet gear $\textcircled{P}$ simultaneously rotates cw about axes C and O. This motion resembles that of a planet moving about its sun, as in our solar system—hence the use of the terms *planetary gear train, planet gear,* and *sun gear.*

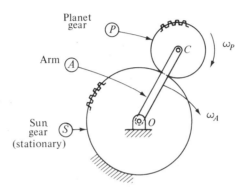

Fig. 8.14. Simple planetary gear train.

Figure 8.15(a) shows a popular type of planetary gear train where several input-output combinations are possible. One possibility is that input motions are applied to the sun gear and arm and an output motion is obtained from the ring gear $\textcircled{R}$ (internal gear). To analyze the motions of this planetary, as well as other planetaries, we shall derive a general equation.

Using Fig. 8.15(a) as reference, the relative angular velocity of sun gear $\textcircled{S}$ and ring gear $\textcircled{R}$ with respect to the arm can be expressed as

$$\omega_{S/A} = \omega_{S/E} \rightarrow \omega_{A/E} \quad \text{and} \quad \omega_{R/A} = \omega_{R/E} \rightarrow \omega_{A/E}$$

These equations are very similar in form to the relative linear velocity equations discussed in Sect. 4.11. Assuming that $\textcircled{S}$, $\textcircled{R}$, and $\textcircled{A}$ move in the same plane or parallel planes, as is the case for many planetaries, the vector equations can be changed to algebraic equations. Since most of the terms in the relative angular velocity equations are understood to be relative to the frame or earth $\textcircled{E}$, the subscripts E will be dropped. The relative angular velocity equations now appear as

$$\omega_{S/A} = \omega_S - \omega_A \quad \text{and} \quad \omega_{R/A} = \omega_R - \omega_A$$

Dividing $\omega_{S/A}$ by $\omega_{R/A}$, we obtain

$$\frac{\omega_S - \omega_A}{\omega_R - \omega_A} = \frac{\omega_{S/A}}{\omega_{R/A}} \tag{1}$$

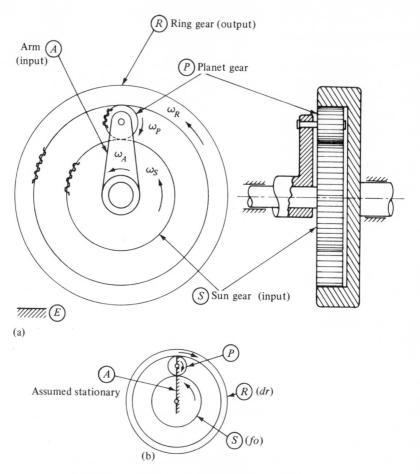

Fig. 8.15. (a) Planetary gear train with inputs to the sun and arm; output is on the ring gear. (b) The planetary shown in (a) with the arm considered stationary.

To evaluate the angular velocity ratio $\omega_{S/A}/\omega_{R/A}$, imagine yourself positioned on the arm $\widehat{A}$. From arm $\widehat{A}$ the planetary appears as an ordinary gear train. The same effect can be accomplished by imagining the arm to be stationary. To visualize this effect, Fig. 8.15(b) is provided to show the appearance of the planetary when the arm is assumed stationary.

All angular velocities in Eq. (1) are either plus or minus values. The sign for ratio $\omega_{S/A}/\omega_{R/A}$ is determined by noting the rotations of $\widehat{S}$ and $\widehat{R}$ with the arm considered stationary. If both gears rotate in the same direction, the ratio sign is plus. If they rotate opposite to one another, the ratio is negative. For this particular planetary, the ratio sign can be readily determined from Fig.

8.15(b). Since $\textcircled{S}$ and $\textcircled{R}$ rotate in opposite directions, the ratio is negative. Be very careful; do not compare the absolute rotations of $\textcircled{S}$ and $\textcircled{R}$ shown in Fig. 8.15(a).

The numerical value for the speed ratio $\omega_{S/A}/\omega_{R/A}$ is determined from Fig. 8.15(b). Considering the planetary arm stationary, gears $\textcircled{R}$, $\textcircled{P}$, and $\textcircled{S}$ appear as an ordinary gear train; thus

$$\frac{\omega_{S/A}}{\omega_{R/A}} = -\frac{N_R(N_P)}{N_P(N_S)} = -\frac{N_R}{N_S}$$

Substituting $-N_R/N_S$ into Eq. (1), we obtain the planetary equation for Fig. 8.15(a):

$$\frac{\omega_S - \omega_A}{\omega_R - \omega_A} = -\frac{N_R}{N_S} \tag{2}$$

To establish a general form for Eq. (2), relabel the following parts; let $\textcircled{S} = \text{fo}$, $\textcircled{R} = \text{dr}$, and $\textcircled{A} = \text{arm}$. Do not interchange or confuse the label fo and dr with the terms output and input. The tooth ratio for the given planetary in terms of the new labels is $-N_R/N_S = -N_{dr}/N_{fo}$. Since not all planetaries can be simply expressed as $-N_{dr}/N_{fo}$, use the gear train tooth ratio $\times N_{dr}/\times N_{fo}$. Substituting the new designations into Eq. (2), we obtain the general form for the planetary equation:

$$\frac{\omega_{fo} - \omega_{arm}}{\omega_{dr} - \omega_{arm}} = \frac{\times N_{dr}}{\times N_{fo}} \tag{8.4}$$

The obtaining of a correct answer from the general planetary equation requires the proper handling of several details. The initial step is to identify the fo and dr gears. Considering the planetary arm stationary, label either end gear in the train as dr; the other end gear is labeled fo. Do not be confused by the fact that the absolute angular velocity of an end gear may be zero. Simply regard fo and dr as labels.

The sign of the tooth ratio, which appears on the right-hand side of Eq. (8.4), is independent of the assumed plus direction for the absolute angular velocities. To determine the sign of the tooth ratio, imagine the planetary arm to be stationary. Now, if the fo gear rotates in the same direction as the dr gear, the ratio is plus; if in the opposite direction, the ratio is negative.

Absolute angular velocities are relative to the machine frame or earth. Since absolute angular velocity can be cw or ccw, a decision must be made as to what direction constitutes plus. To reduce the chance of mathematical error, assume the input direction, to the planetary, to be plus. In the case of two opposite inputs, obviously only one input can be chosen as plus. In Fig. 8.15(a), input rotations to $\textcircled{S}$ and $\textcircled{A}$ are ccw; thus it would be wise to assume ccw as plus.

It is interesting to note that the general planetary equation

$$\frac{\omega_{fo} - \omega_{arm}}{\omega_{dr} - \omega_{arm}} = \frac{\times N_{dr}}{\times N_{fo}}$$

reduces to the ordinary gear train equation

$$\frac{\omega_{fo}}{\omega_{dr}} = \frac{\times N_{dr}}{\times N_{fo}}$$

when $\omega_{arm} = 0$.

8.9 Planetary Gear Train

Planetary gear trains are frequently called epicyclic gear trains.

The general planetary equation will be applied only to a planetary; therefore if a gear train consists of a planetary and an ordinary gear train, it will be necessary to analyze them separately.

A large speed reduction or increase can be obtained from a relatively small planetary unit. For very large speed changes, two or more planetaries can be connected in series.

To set up the general planetary equation, it is recommended that a systematic approach be adopted. One such procedure is

STEP 1. Write down the general planetary equation [Eq. (8.4)]. Draw a vertical line separating the left-hand side (relative to the frame) from the right-hand side (relative to the arm).

STEP 2. Start on the right-hand side of the equation; with the arm assumed stationary, choose a dr gear. Now determine the tooth ratio $\times N_{dr}/\times N_{fo}$.

STEP 3. Now move to the left-hand side of the equation. Assume a plus direction of rotation. Change subscripts for ω_{dr}, ω_{fo}, and ω_{arm} to letters which correspond to the problem. Where appropriate, assign numerical values to the angular velocities.

STEP 4. Substitute into the general planetary equation and solve.

The following examples illustrate how the general planetary equation is applied.

EXAMPLE 8.6

Given: A planetary gear train (Fig. 8.16) has two inputs. The sun gear and arm rotate at 500 and 300 rpm ccw, respectively.

Determine: Rpm and direction of rotation for the ring gear.

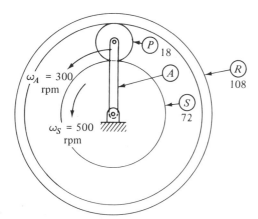

Fig. 8.16.

Solution. Following the four-step procedure just outlined (start on the right-hand side of the planetary equation):

$$\frac{\omega_{fo} - \omega_{arm}}{\omega_{dr} - \omega_{arm}} = \frac{\times N_{dr}}{\times N_{fo}}$$

Relative to frame　　　　　　Assume arm stationary
Assume ccw +　　　　　　　Assume dr gear is ®

$$\omega_{dr} = \omega_R = ?$$　　　　$$\frac{\times N_{dr}}{\times N_{fo}} = -\frac{108}{18}\left(\frac{18}{72}\right) = -\frac{3}{2}$$

$$\omega_{fo} = \omega_S = 500$$

$$\omega_{arm} = \omega_A = 300$$

Substitute values into the equation

$$\frac{500 - 300}{\omega_R - 300} = -\frac{3}{2}$$

$$1000 - 600 = -3\omega_R + 900$$

$$\omega_R = 166.7 \text{ rpm}$$

A plus answer indicates that ® rotates ccw.

EXAMPLE 8.7

Given: The boom rotating mechanism for an aerial device is shown in Fig. 8.17. A hydraulic motor drives the sun gear at 17.5 rev/sec. Ring gear ® is stationary.

Determine: The rev/sec for the boom gear.

Solution. In this example a planetary is connected to an ordinary gear train. Here the procedure is to analyze the planetary and then the ordinary gear train.
　Start on right-hand side of the planetary equation:

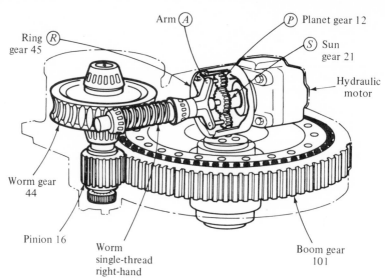

Arm (A) (P) Planet gear 12

Ring (R) gear 45

(S) Sun gear 21

Hydraulic motor

Worm gear 44

Pinion 16

Worm single-thread right-hand

Boom gear 101

Fig. 8.17. *Aerial boom drive train.* (Courtesy of Pitman Mfg. Co.)

$$\frac{\omega_{fo} - \omega_{arm}}{\omega_{dr} - \omega_{arm}} = \frac{\times N_{dr}}{\times N_{fo}}$$

Relative to frame
Assume input $+$

$$\omega_{dr} = \omega_R = 0$$

$$\omega_{fo} = \omega_S = 17.5$$

$$\omega_{arm} = \omega_A = \ ?$$

Assume arm stationary
Assume dr gear is (R)

$$\frac{\times N_{dr}}{\times N_{fo}} = -\frac{45}{12}\left(\frac{12}{21}\right) = -\frac{15}{7}$$

Substitute into the planetary equation:

$$\frac{17.5 - \omega_A}{0 - \omega_A} = -\frac{15}{7}$$

$$122.5 - 7\omega_A = 15\omega_A$$

$$\omega_A = \frac{122.5}{22} \text{ rev/sec}$$

The plus answer indicates that the output or arm rotates in the same direction as the input.

In the ordinary gear train, $\omega_{dr} = 122.5/22$ rev/sec:

$$\omega_{fo} = \omega_{dr}\left(\frac{\times N_{dr}}{\times N_{fo}}\right) = \frac{122.5}{22}\left[\frac{1(16)}{44(101)}\right] = .02 \text{ rev/sec}$$

Thus the boom gear rotates at .02 rev/sec. The overall speed ratio for the entire gear train is $.02/17.5 = 1/875$, or a speed reduction of 875.

EXAMPLE 8.8

Given: Figure 8.18 shows a reverted planetary gear train B, C, D, and E. Planet gears C and D are keyed to the same shaft. The 45- and 30-tooth cluster gears rotate at 1000 rpm. Arm A is connected to the 45-tooth gear.

Determine: Overall speed ratio (output/input).

Solution. The planetary gear train is B, C, D, and E, where $\omega_A = 1000$ rpm and $\omega_B = 500$ rpm:

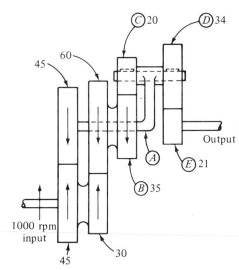

Fig. 8.18. *Reverted planetary and ordinary gear trains.*

$$\frac{\omega_{fo} - \omega_{arm}}{\omega_{dr} - \omega_{arm}} = \frac{\times N_{ar}}{\times N_{fo}}$$

<table>
<tr><td>

Relative to frame
Assume ccw +
  (viewed from right)

$\omega_{dr} = \omega_B = 500$

$\omega_{fo} = \omega_E = \, ?$

$\omega_{arm} = \omega_A = 1000$

</td><td>

Assume arm stationary
Assume dr gear is $\textcircled{B}$

$\dfrac{\times N_{dr}}{\times N_{fo}} = \dfrac{35}{20}\left(\dfrac{34}{21}\right) = \dfrac{17}{6}$

</td></tr>
</table>

$$\frac{\omega_E - 1000}{500 - 1000} = \frac{17}{6}$$

$$6\omega_E - 6000 = -8500$$

$$\omega_E = -416.7 \text{ rpm}$$

Gear $\textcircled{E}$ rotates cw when viewed from the right. The overall speed ratio is

$$\frac{\text{output}}{\text{input}} = \frac{416.7}{1000} = .4167$$

Note that both input (1000 rpm) and output (416.7 rpm) rotate in the same direction.

EXAMPLE 8.9

Given: Figure 8.19 shows a phase-shifting differential. Input rotates at 1500 rpm.

Determine: Output rpm; consider gear $\textcircled{D}$ stationary.

Solution. This bevel gear differential consists of two spider gears or planet gears which rotate freely about the spider arms while pivoting about the output axis. When kinematically analyzing a planetary, it is usually convenient to consider only one of the spider gears in the gear train; thus gear $\textcircled{S}$ will be disregarded.

The phase angle between the input and output shafts can be adjusted by moving gear $\textcircled{D}$ via worm and worm gear. This adjustment can take place while the mechanism is running. Angular drift is eliminated by making the worm gearset self-locking.

Analyzing planetary train $\textcircled{B}$, $\textcircled{C}$, and $\textcircled{D}$ with $\textcircled{D}$ being stationary,

$$\frac{\omega_{fo} - \omega_{arm}}{\omega_{dr} - \omega_{arm}} = \frac{\times N_{dr}}{\times N_{fo}}$$

<table>
<tr><td>

Relative to frame
Assume cw +
  (viewed from right)

$\omega_{dr} = \omega_D = 0$

$\omega_{fo} = \omega_B = 1500$

$\omega_{arm} = \omega_A = \, ?$

</td><td>

Assume arm stationary
Assume dr gear is $\textcircled{D}$

$\dfrac{\times N_{dr}}{\times N_{fo}} = -\dfrac{50}{32}\left(\dfrac{32}{50}\right) = -1$

</td></tr>
</table>

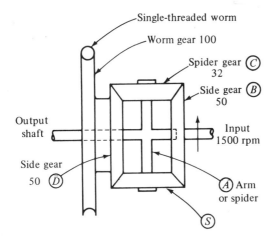

Fig. 8.19. *Phase-shifting differential transmission.* (Courtesy of Fairchild Hiller Corp.)

$$\frac{1500 - \omega_A}{0 - \omega_A} = -1$$

$$\omega_A = 750 \text{ rpm}$$

The output rotates in the same direction as the input.

Let us determine the angular shift between input and output shafts for one complete rotation of the worm. If the worm rotates 1 rev, the worm gear and D rotate $\frac{1}{100}$ rev, or 3.6 deg. Using the general planetary equation with the input shaft stationary (gear B),

$$\frac{\omega_{fo} - \omega_{arm}}{\omega_{dr} - \omega_{arm}} = \frac{\times N_{dr}}{\times N_{fo}}$$

Relative to frame	Assume arm stationary
Assume rotation $+$	Assume dr gear is $\textcircled{D}$

$$\omega_{dr} = \omega_D = 3.6$$
$$\omega_{fo} = \omega_B = 0 \qquad \frac{\times N_{dr}}{\times N_{fo}} = -1$$
$$\omega_{arm} = \omega_A = \,?$$

$$\frac{0 - \omega_A}{3.6 - \omega_A} = -1$$

$$\omega_A = 1.8°$$

Therefore 1 rev of the worm results in an angular change of 1.8 deg between the input and output shafts. What would be the number of revolutions required for a 9-deg shift between the input and output shafts? Using the information that there is a 1.8-deg shift for 1 rev of the worm, we have 9 deg (1 rev/1.8 deg) = 5 rev of the worm is required.

8.10 Speed Changers— Planetary Gear Type

In a planetary gear train, speeds can be changed without shifting gears with the aid of clutches and brakes.

Figure 8.20 shows a two-speed planetary gear train with one clutch and one brake. The input shaft is keyed to the sun gear and splined to the clutch. The output shaft is fastened to the planet carrier, and the ring gear is part of the brake drum.

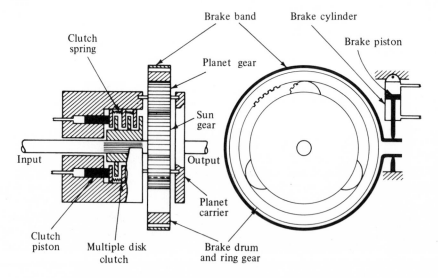

Fig. 8.20. *Simplified drawing of a two-speed planetary transmission.*

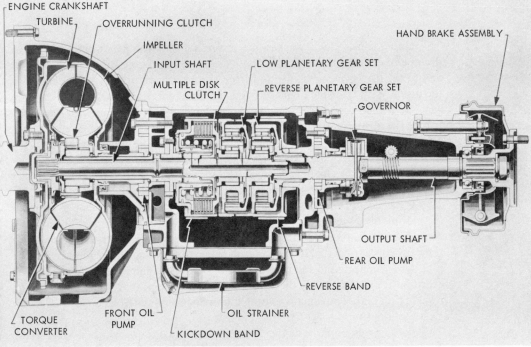

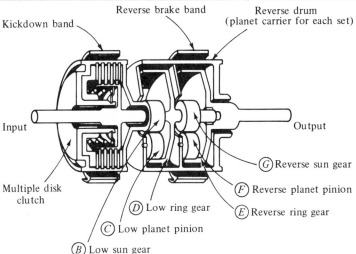

Fig. 8.21. *Powerflite transmission.* (Courtesy of Dodge Division of Chrysler Motors Corp.)

A speed reduction is obtained by pressurizing the top of the brake cylinder. This causes the brake piston to move downward, tightening the brake band around the drum. Since the brake band is held stationary, the drum and ring gear will also be held stationary. The output rotation obtained is in the same direction as the input but at a reduced speed. Note that the clutch is not engaged for this speed.

To obtain high speed or direct drive, the multiple disk clutch is engaged.

This clutch consists of a washer-like piston which when pressurized moves the driver and driven disks together. These disks are alternately splined to the driver and to the driven member. Motion from the input is transmitted by friction via the disks. With the clutch engaged and the brake off, the planet carrier and sun gear are locked together. For a simple planetary system such as this, when any two of the three gears are locked together, the entire system rotates as one unit. Thus the output turns at the same speed as the input. The clutch springs are used to separate the disks once the fluid is depressurized.

For neutral, the clutch is disengaged and the brake is off. The input shaft can rotate but the output shaft is not positively driven.

An excellent example of an automatic planetary transmission can be found in the automobile. Figure 8.21 shows the Chrysler Motor Powerflite transmission. This is an automatic two-forward and one-reverse speed transmission. The transmission consists of two simple planetary sets where the planet carrier for each set is connected to the reverse drum. The input shaft is part of the low ring gear and is also splined to the reverse sun gear. The output shaft is part of the reverse ring gear. The function of the multiple disk clutch is to engage or disengage the input motion to the low sun gear.

Power flow is from the engine crankshaft through the torque converter to the input shaft of the planetary transmission. The torque converter is a hydraulic unit which transmits power without shock or vibration. Torque can be increased or multiplied in the converter.

Four positons of the transmission will be discussed: low, drive, reverse, and neutral.

Low. The kickdown band is applied; this holds the low sun gear stationary. The clutch is disengaged and the reverse brake band is off. To calculate the speed reduction, assume that the input rotates 1 rev, and use the schematic shown in Fig. 8.22(a). Take careful note that the planet pinions $\textcircled{C}$ and $\textcircled{F}$ are not keyed together; that is, they do not necessarily rotate at the same speed. The planetary equation must be applied twice. First, solve low planetary train $\textcircled{B}$, $\textcircled{C}$, and $\textcircled{D}$:

$$\frac{\omega_{fo} - \omega_{arm}}{\omega_{dr} - \omega_{arm}} = \frac{\times N_{dr}}{\times N_{fo}}$$

Relative to frame | Assume arm stationary
Assume ↑ + | Assume dr is $\textcircled{D}$

$$\omega_{dr} = \omega_D = 1$$

$$\omega_{fo} = \omega_B = 0$$

$$\omega_{arm} = \omega_A = \,?$$

$$\frac{\times N_{dr}}{\times N_{fo}} = -\frac{55(16)}{16(23)} = -\frac{55}{23}$$

$$\frac{0 - \omega_A}{1 - \omega_A} = -\frac{55}{23}$$

$$\omega_A = .705 \text{ rev}$$

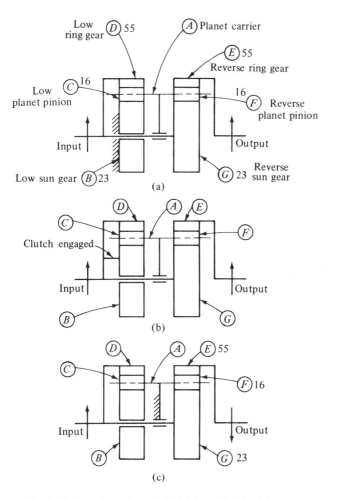

Fig. 8.22. Powerflite schematic for (a) low, (b) drive, (c) reverse speed.

The arm or carrier rotates in the same direction as the input. Now apply the planetary equation to reverse planetary Ⓔ, Ⓕ, and Ⓖ where the carrier rotates .705 rev and the reverse sun gear Ⓖ rotates in the same direction with an input motion of 1 rev:

Relative to frame
Assume ↑ +

$$\omega_{dr} = \omega_E = ?$$

$$\omega_{fo} = \omega_G = 1$$

$$\omega_{arm} = \omega_A = .705$$

Assume arm stationary
Assume dr is Ⓔ

$$\frac{\times N_{dr}}{\times N_{fo}} = -\frac{55}{23}$$

$$\frac{1 - .705}{\omega_E - .705} = -\frac{55}{23}$$

$$\omega_E = .582 \text{ rev}$$

For a 1-rev input, the output moves .582 rev in the same direction. The input to output speed is $1/.582 = 1.72$.

Drive. The low sun gear Ⓑ is locked to the input by engaging the clutch. Both bands are off. The low planetary set rotates as one unit because Ⓓ and Ⓑ are rotating at the same speed [refer to Fig. 8.22(b)]. Also, the reverse planetary set rotates as one unit because the planet carrier Ⓐ and reverse sun gear Ⓖ are rotating at the same speed. Thus it can be said that the entire transmission rotates as one solid unit. The output turns at the same speed as the input.

Reverse. The reverse band is applied; this holds the planet carriers stationary. The clutch and kickdown band are disengaged. With the carriers stationary, the low and reverse sets now appear as simple gear trains [see Fig. 8.22(c)]. Since the low gear set cannot transmit power, we shall analyze only the reverse gear set. Using an input of 1 rev,

$$\omega_{fo} = \omega_{dr}\left(\frac{\times N_{dr}}{\times N_{fo}}\right)$$

$$\omega_E = \left(\frac{23}{16}\right)\left(\frac{16}{55}\right) = .418 \text{ rev}$$

Here the output rotates opposite to the input direction.

Neutral. The clutch is disengaged and both bands are off.

8.11 Design of a Planetary Gear Train

This section is concerned with the design of a basic planetary gear train which consists of a ring gear, a planet gear (or gears), a sun gear, and a planet carrier (Fig. 8.23).

It was established in Sect. 8.8 that the equation for a basic planetary gear train is

$$\frac{\omega_S - \omega_A}{\omega_R - \omega_A} = -\frac{N_R}{N_S}$$

From this equation we obtain

$$N_S\omega_S + N_R\omega_R = \omega_A(N_S + N_R) \qquad (1)$$

Two output-input relationships can be established from Equation (1) by holding the ring gear stationary ($\omega_R = 0$). Two other output-input equations can be determined by holding the sun gear stationary ($\omega_S = 0$). These equations are easily derived and for convenience are listed in Fig. 8.23. It should be noted that

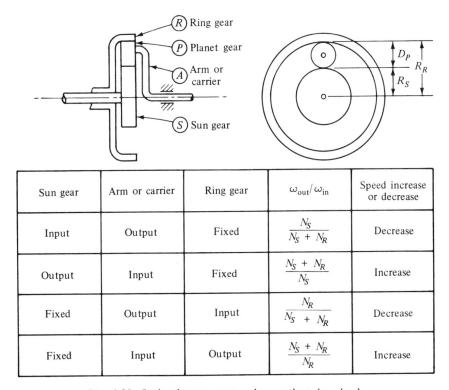

Sun gear	Arm or carrier	Ring gear	ω_{out}/ω_{in}	Speed increase or decrease
Input	Output	Fixed	$\dfrac{N_S}{N_S + N_R}$	Decrease
Output	Input	Fixed	$\dfrac{N_S + N_R}{N_S}$	Increase
Fixed	Output	Input	$\dfrac{N_R}{N_S + N_R}$	Decrease
Fixed	Input	Output	$\dfrac{N_S + N_R}{N_R}$	Increase

Fig. 8.23. *Basic planetary gear train equations* (ω_{out}/ω_{in}).

the input and output of a basic planetary gear train rotate in the same direction.

Referring to the equations shown in Fig. 8.23, for a speed reduction the output is always out of the arm, and for a speed increase the input is always into the arm. Since the largest speed reduction and increase, for a given N_S and N_R, is obtained by keeping the ring gear stationary, we shall usually use the first two equations from Fig. 8.23.

For speed reduction,

$$\frac{\omega_{out}}{\omega_{in}} = \frac{N_S}{N_S + N_R} \tag{8.5}$$

where $\omega_{out} = \omega_{arm}$, $\omega_{in} = \omega_{sun}$, and the ring gear is stationary.

For speed increase,

$$\frac{\omega_{out}}{\omega_{in}} = \frac{N_S + N_R}{N_S} \tag{8.6}$$

where $\omega_{out} = \omega_{sun}$, $\omega_{in} = \omega_{arm}$, and the ring gear is stationary.

For a given output-input ratio, the tooth numbers N_S and N_R can be established from Eqs. (8.5) or (8.6). The question now occurs, what is the number of teeth on the planet gear. To determine the number of teeth on the planet gear, refer to Fig. 8.23. Here

$$R_R = R_S + D_P$$

or

$$D_R = D_S + 2D_P$$

For standard spur gears, $D = N/P$; thus

$$\frac{N_R}{P} = \frac{N_S}{P} + \frac{2N_P}{P}$$

Since the diametral pitch P is the same for all gears, the equation can be simplified to

$$N_R = N_S + 2N_P$$

Therefore

$$N_P = \frac{N_R - N_S}{2} \tag{8.7}$$

This relationship is also true for helical gears.

Several equally spaced planet gears are often employed to increase the load-carrying capacity of the planetary, as well as to balance centrifugal forces. For the gears to fit properly, with equal spacing,

$$\frac{N_S + N_R}{M} = \text{whole number} \tag{8.8}$$

where M is the number of equally spaced planet gears. This equation does not consider the possibility of interference between adjacent planet gears.

EXAMPLE 8.10

Given: The output shaft of a basic planetary gear train is to rotate at 100 rpm. Input rotates at 450 rpm.

Determine: Number of teeth on each gear in the planetary. Check to see if three equally spaced planet gears will fit properly.

Solution. This is a speed-reduction problem. The input is into the sun gear, the output is from the arm, and the ring gear is stationary. Use Eq. (8.5):

$$\frac{\omega_{\text{out}}}{\omega_{\text{in}}} = \frac{N_S}{N_S + N_R}$$

$$\frac{100}{450} = \frac{N_S}{N_S + N_R}$$

$$\frac{N_S + N_R}{N_S} = \frac{450}{100}$$

$$\frac{N_R}{N_S} = 3.5$$

Let $N_R = 84$ and $N_S = 24$. To determine N_P, use Eq. (8.7):

$$N_P = \frac{N_R - N_S}{2} = \frac{84 - 24}{2} = 30$$

To check for three equally spaced planet gears, use Eq. (8.8):

$$\frac{N_S + N_R}{M} = \frac{24 + 84}{3} = 36$$

Since the result is a whole number, three planet gears can be equally spaced around the sun gear.

For a table of gear combinations, the reader is referred to H. Reed Langdon, "Planetary Gear Ratios," *Machine Design*, April 28, 1960.

PROBLEMS

8.1 The output gear of a simple gear train (Fig. P8.1) rotates at 100 rpm cw. Determine the input rpm.

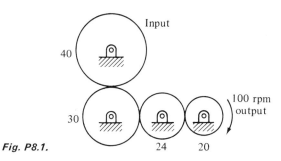

Fig. P8.1.

8.2 Using linear velocity vectors, such as V_X, V_Y, and V_Z, show that the gear train speed ratio for Fig. P8.2 is $\omega_D/\omega_A = N_A/N_D$.

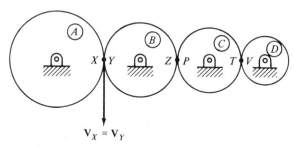

$$\mathbf{V}_X = \mathbf{V}_Y$$

Fig. P8.2.

8.3 Figure P8.3 shows a cutaway view of a two-speed portable electric drill. There are 7 teeth on the armature pinion, 39 teeth on the intermediate helical gear, 12 teeth on the intermediate spur gear, and 53 teeth on the spindle gear. Low speed is obtained by changing the frequency of the supply voltage to the ac motor. Determine the spindle rpm's if the armature speeds are 24,600 and 12,300 rpm.

Fig. P8.3. (Courtesy of Portable Electric Tools.)

8.4 Determine the gear train speed ratio for the servomechanism shown in Fig. P8.4.

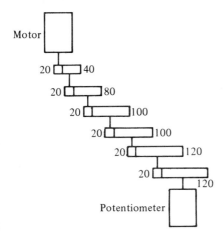

Fig. P8.4.

8.5 The front wheels of a toy truck are driven by a compound gear train (Fig. P8.5). Gears B and C, D and E, and F and G are compounded. Gear H is keyed to the front wheel shaft. Assume that the 1.5-volt dc motor rotates at 5000 rpm. Determine (a) the wheel rpm and (b) the distance the truck will move in 10 sec.

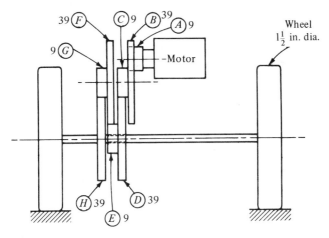

Fig. P8.5.

8.6 Figure P8.6 shows the paper drive mechanism of a temperature recorder. The mechanism is driven by a synchronous motor which rotates at 1800 rpm. The sides of the paper strip are driven by a 12-tooth drum. The distance between adjacent holes on the paper is .5 in. Determine (a) the paper speed (ft/min) and (b) the direction of rotation for the synchronous motor so that the paper moves as shown in Fig. P8.6.

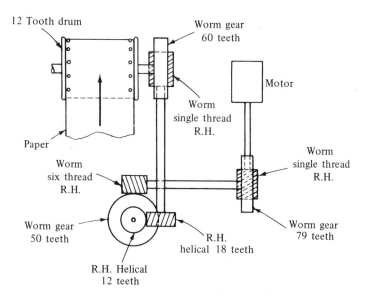

Fig. P8.6.

8.7 Determine the output rpm and the direction of rotation for the 20-tooth gear shown in Fig. P8.7. The input rotates at 2000 rpm. Assume that the belts do not slip.

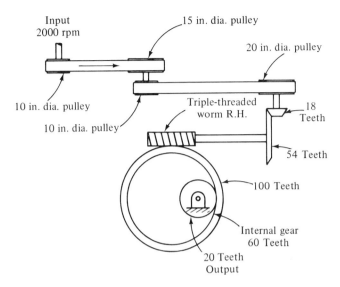

Fig. P8.7.

8.8 A selective speed reducer is shown in Fig. P8.8. Speed is changed by engaging and disengaging clutches on the input and output shafts. Each clutch is fastened rigidly to an adjacent gear. The input shaft rotates at 1000 rpm. Determine the output speeds.

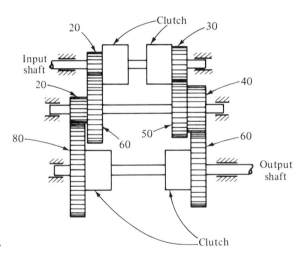

Fig. P8.8.

8.9 Nine ratios can be obtained from the manually operated geared transmission shown in Fig. P8.9. Determine these ratios. Note that the transmission utilizes two shift levers.

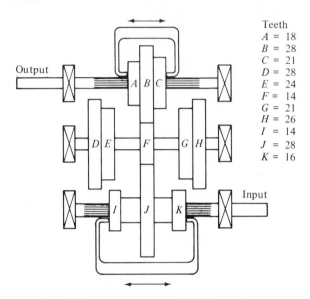

Teeth
$A = 18$
$B = 28$
$C = 21$
$D = 28$
$E = 24$
$F = 14$
$G = 21$
$H = 26$
$I = 14$
$J = 28$
$K = 16$

Fig. P8.9. (Courtesy of Turner Uni-Drive Co.)

8.10 Figure P8.10 shows a dial-operated six-speed gearmotor. A shift cam, driven by a selector gear, moves the sliding gears into and out of mesh with the 55-tooth center gear. There are six equally spaced identical sliding gears mounted around the center gear (only two are shown). Determine the output speeds if the input shaft rotates at 100 rpm. Only three of the six gear trains are shown.

Fig. P8.10. (Courtesy of Geartronics Corporation.)

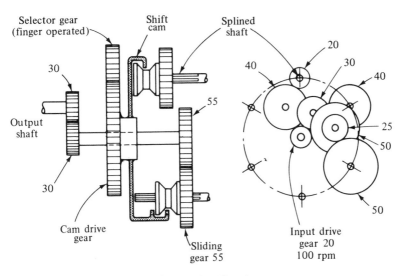

Fig. P8.10. (Cont.)

8.11 For the five-speed transmission (synchronized in second, third, fourth, and fifth) shown in Fig. P8.11, determine the inverse of the gear train speed ratio for all speeds, including reverse.

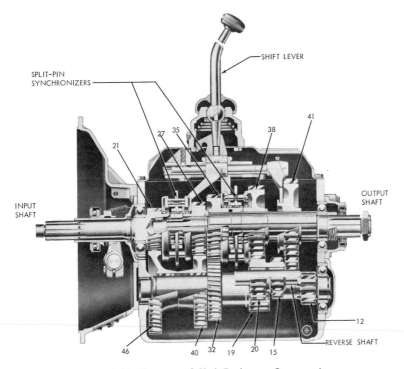

Fig. P8.11. (Courtesy of Clark Equipment Company.)

8.12 Figure P8.12 shows a fully synchronized three-speed forward manual transmission. Determine the inverse of the gear train speed ratio for all speeds, including reverse. The 15-tooth cluster gear (hidden from view) is in constant mesh with both the 32-tooth low gear and the 17-tooth reverse idler gear. The 28-, 20-, and 15-tooth gears are keyed to the same shaft.

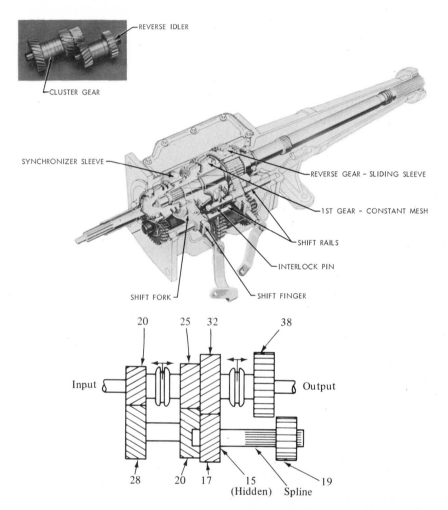

Fig. P8.12. (Courtesy of Ford Motor Company.)

8.13 Design a compound gear train for a speed ratio of 1:20. No gear is to have less than 20 teeth or more than 120 teeth. The output gear is to rotate opposite to the input gear. Sketch the completed gear train.

8.14 Design a compound gear train for a speed ratio of 2:25. No gear is to have less than 18 teeth or more than 54 teeth.

8.15 Design a compound gear train for a speed ratio of 8.25:1. No gear is to have less than 16 teeth or more than 96 teeth.

8.16 A speed reducer is to convert 2100 rpm into 90 rpm. (a) Design a compound gear train to accomplish this. Use a minimum of 15 teeth and a maximum of 120 teeth. (b) Design a geared unit, other than a compound gear train, to have an output of 90 rpm.

8.17 Design a gear train for a speed ratio of .2944 ± .0001. Use a minimum of 16 teeth and a maximum of 96 teeth.

8.18 Design a reverted gear train for a speed ratio of 1:12. Use a minimum of 16 teeth and a maximum of 120 teeth. All gears have the same pitch.

8.19 Design a reverted speed increaser to have a ratio of 7.5:1. Use neither less than 18 teeth nor more than 72 teeth/gear. All gears have the same pitch.

8.20 Design a reverted spur gear train where the center distance is approximately 2 in. The diametral pitch for all gears is to be 32. The minimum number of teeth per gear is 20 and the maximum number is 120. The required speed ratio is 1:10.

8.21 In Fig. P8.21, select spur gears from a gear catalog for a speed reduction of 45. Use the same diametral pitch for all spur gears. The diametral pitch is limited to 48, 64, or 72. Calculate the center distance between parallel shafts.

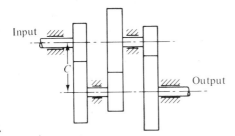

Fig. P8.21.

8.22 For the planetary train shown in Fig. P8.22, determine the angular velocity of the arm and planet gear. The ring gear rotates at 500 rpm cw, and the sun gear rotates at 750 rpm ccw.

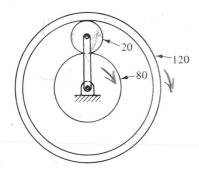

Fig. P8.22.

8.23 The ring gear in Fig. P8.22 rotates at 500 rpm ccw, and the sun gear rotates at 750 rpm ccw. Determine the arm rpm.

8.24 In Fig. P8.24, gear B is stationary. Determine ω_D/ω_A.

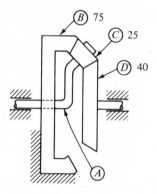

Fig. P8.24.

8.25 Figure P8.25 shows a simplified sketch of a planetary train housed within a pulley case. The casing drives the planet gears C and D; these gears are keyed to the same shaft. Sun gear B is held stationary, and sun gear E is connected to the output shaft. If the motor speed is 1800 rpm, what is the output shaft rpm. Note that there is a 3 : 1 reduction in the V-belt drive.

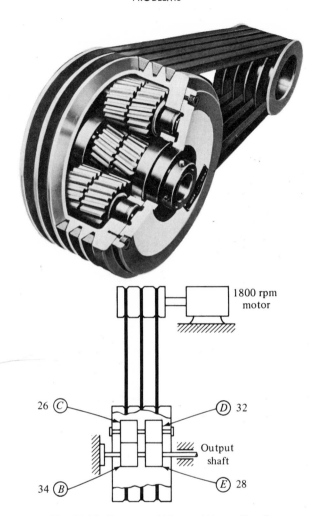

Fig. P8.25. (Courtesy of Plessey Airborne Corp.)

8.26 The torque-meter shown in Fig. P8.26 utilizes a bevel gear differential. A mechanism, not shown in the drawing, is connected to the spider shaft. The spider shaft pivots either toward the reader or away from the reader. The magnitude of this movement is dependent on the amount of torque or angular displacement between the ends of the torsional element. Note that the load is not transmitted through the gears; the torsional element carries the load. Assume that the angular twist between the power end and the load end is 2 deg. What is the angular movement of the spider shaft?

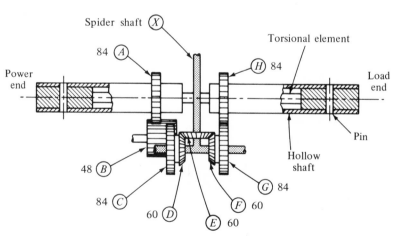

Fig. P8.26. (Courtesy of Power Instruments Inc.)

8.27 Bevel gear differentials (Fig. P8.27) can be used to add and subtract rotations. Mathematically show that when both side gears rotate in the same direction, the spider shaft rotates at half the sum of both speeds.

Fig. P8.27. (Courtesy of Winfred M. Berg. Inc.)

8.28 Figure P8.28 shows a bevel gear differential which connects the rear axles of an automobile. The planet carrier and ring gear are one unit. When making a right-hand turn, the right wheel rotates at 14 rad/sec and the left at 15 rad/sec. Determine the angular velocity of the drive pinion.

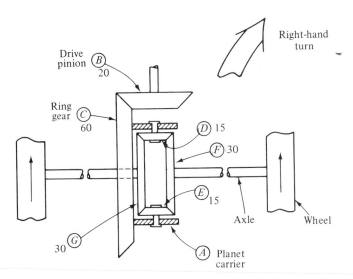

Fig. P8.28.

8.29 In Fig. P8.29, the triple-threaded right-hand worm rotates at 1800 rpm cw. The planetary train consists of gears $Ⓡ$, $Ⓟ$, and $Ⓢ$. Determine the rpm for the output shaft and the planet gear. When solving for the planet gear rotation, use the equation $(\omega_P - \omega_A)/(\omega_S - \omega_A) = \omega_{P/A}/\omega_{S/A}$.

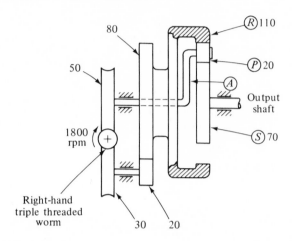

Fig. P8.29.

8.30 The planetary gear train shown in Fig. P8.30 requires the double application of the planetary equation. Determine the angular velocity for gear $Ⓔ$ when gear $Ⓑ$ rotates at 1800 rpm. Internal gear $Ⓕ$ is stationary. First, use the equation $(\omega_B - \omega_A)/(\omega_F - \omega_A) = \omega_{B/A}/\omega_{F/A}$, and then use the equation $(\omega_E - \omega_A)/(\omega_F - \omega_A) = \omega_{E/A}/\omega_{F/A}$. Be careful evaluating the tooth ratios.

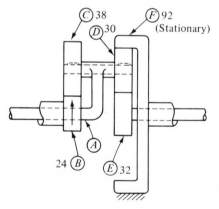

Fig. P8.30.

8.31 Figure P8.31 is a reduction gear train used in a gas turbine engine. The planetary train is $Ⓑ$, $Ⓒ$, and $Ⓓ$. The simple gear train is $Ⓔ$, $Ⓕ$, and $Ⓖ$. Gear $Ⓓ$ is coupled to gear $Ⓔ$. The output is from both $Ⓐ$ and $Ⓖ$. Determine the output rpm for an input of 19,350 rpm. When assigning algebraic signs to the absolute angular velocities, note that gear $Ⓑ$ rotates opposite to gear $Ⓓ$ and $Ⓔ$. Also note that $|\omega_E| = \omega_G(\frac{137}{55})$.

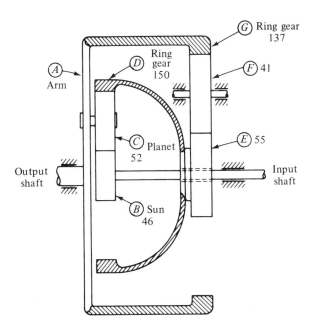

Fig. P8.31. (Courtesy of Avco Lycoming Div.)

8.32 A simple planetary train consists of a sun gear, planet gear, ring gear, and planet carrier. Complete the table shown in Fig. P8.32; use the symbols N_S, N_P, and N_R to represent the teeth on the sun, planet, and ring gears, respectively.

Condition	Sun gear	Arm or carrier	Ring gear	$\dfrac{\omega_{in}}{\omega_{out}}$	Speed increase or decrease
1	Input	Output	Fixed		
2	Output	Input	Fixed		
3	Fixed	Output	Input		
4	Fixed	Input	Output		
5	Input	Fixed	Output		
6	Output	Fixed	Input		
7	Entire train locked together				

Fig. P8.32.

8.33 For Fig. 8.20, assume that the sun gear has 60 teeth, that the planet gear has 21 teeth, and that the ring gear has 102 teeth. Determine all possible output speeds for an input of 1000 rpm.

8.34 Figure P8.34 shows part of a tractor planetary transmission. Clutches C1 and C2 direct the power from the engine to the inner or outer input shaft. Hydraulic brakes B1 and B2 are used to hold the ring gears stationary. Output motion is obtained from the planet carrier. Planet gears 33 and 27 are keyed to the same shaft. Calculate five possible output speeds for an engine speed of 2000 rpm. For example, one such output speed is obtained by engaging clutch C1 and applying brake B1.

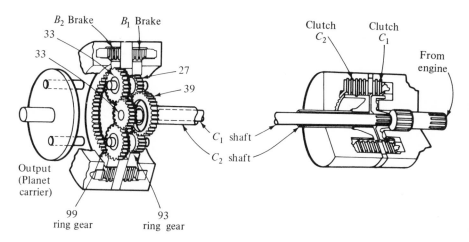

Fig. P8.34. (Courtesy of Deere and Company.)

8.35 Design a basic planetary gear train (Fig. 8.23) for a speed reduction of 5. Do not use less than 18 teeth.

8.36 Design a basic planetary gear train (Fig. 8.23) with three equally spaced planet gears for an output speed of 2000 rpm. The input is 500 rpm.

8.37 Design a basic planetary gear train (Fig. 8.23) with two equally spaced planet spur gears for an output-input ratio of 1:3.50. The ring gear pitch diameter is to be approximately 3 in.; the diametral pitch is 24.

8.38 Design a basic planetary gear train (Fig. 8.23) with three equally spaced planet gears for an input of 1600 rpm and an output of 1000 rpm. The sun gear is to be held stationary.

8.39 Design a two-stage planetary gear train which consists of two basic planetary trains (Fig. 8.23) in series. The overall output-input ratio is 1:32.

Index

V = Vel. ft/min.

D = pitch dia ft. $V = \pi D n$

n = rpm